INTENDED
EVOLUTION

HOW SELECTION *of* INTELLIGENCE

GUIDES LIFE FORWARD

Dongxun Zhang and Bob Zhang

Edited by David Kincade

RIVER GROVE
BOOKS

This book is intended as a reference volume only, not as a medical manual. The information given here is designed to help you make informed decisions about your health. It is not intended as a substitute for any treatment that may have been prescribed by your doctor. If you suspect that you have a medical problem, you should seek competent medical help. You should not begin a new health regimen without first consulting a medical professional.

Published by River Grove Books
Austin, TX
www.rivergrovebooks.com

Distributed by River Grove Books

For ordering information or special discounts for bulk purchases, please contact River Grove Books at PO Box 91869, Austin, TX 78709, 512.891.6100.

Design and composition by Greenleaf Book Group
Cover design by Greenleaf Book Group
Cover photo ©Shutterstock.com/Cessna152

Publisher's Cataloging-in-Publication-data is available.

ISBN: 978-1-63299-018-1

First Edition

Other Edition(s):
eBook ISBN: 978-1-63299-019-8

CONTENTS

—

PREFACE

—

It has been more than twenty years since my theory of intended evolution first came together and I planned to write this book. However, because of work or other matters, it never got done. About ten years ago, I created a fitness program based on this theory, an idea that I called Intended Evolution Fitness, due to my feeling at the time that the need for this particular application of the theory was more important than writing out the pure theory itself. Within the materials for the fitness program was a brief segment on the theory of intended evolution, an idea that drew interest from some of my students and that led to a promise that a book would soon follow. But the book was slow in coming until finally in the last year or so, with the help of David Kincade and Bob Zhang, the project got underway in earnest.

This book is written in the broadest of strokes, as the intention is to state as simply as possible the basic ideas of intended evolution and no more than that. We will not present details of the mechanics of the process nor will we spend a lot of time presenting the evidence to back up our claims. Instead, we discuss our ideas as an underlying framework pertaining to what are mostly generally accepted principles of biology and evolution. Therefore, this book is about an overall "why" rather than "how" processes work. The reason for this is that collecting, organizing and presenting such data requires time and resources that we do not have. Rather than going into great detail about the mechanics of applying the theory of intended evolution across diverse fields of knowledge, we have opted to provide here a brief introduction to a number of concepts, recontexualizing them in light of this theory.

That said, I welcome the accumulation of existing and new work from others that furthers discussion about the usefulness of this framework. At its core, the idea of intended evolution is a simple one: living things have an internal drive that results in interactions with the external environment that are different from those of nonlife in that they lead to intentional changes of the living organisms' internal information, function, and structure.

Furthermore, because of the unique evolutionary history of humans, we can, in a significant way, affect the future of our own evolution going forward.

INTRODUCTION

—

TODAY'S DISCOURSE ON EVOLUTION INEVITABLY REFERS TO NATURAL
selection and genetic variation as the two basic components
of the conceptual framework. The prominent idea, stated
simply, is that random changes in the genome (mutations)
create the variety of individuals from which nature "selects"
survivors. Deviation from this line of thinking is viewed with
great skepticism or even linked with some form of intelligent
design or creationism. This type of dogmatic thinking tends
to be counterproductive to broader analysis and understanding of the evolutionary process. Darwin himself concluded
the introduction to his work, *On the Origin of Species,* with the
following sentence: "I am convinced that Natural Selection
has been the main but not exclusive means of modification."
Therefore, even by Darwin's estimation, there are other factors at work than strictly a selection process.

THREE BASIC FACTORS

Based on what is known about the evolutionary process, we
believe there are three basic factors: external factors, internal

factors, and a combination of internal and external factors. By *external factors*, we mean the environment described as the means of natural selection, according to Darwin. By *internal factors*, we mean an organism's intentional activity or drive and will to survive. With *combination*, we refer to the results of the interface of an organism's intentional activity with the environment that manifests an intelligent way of living, such as behavioral habits. This category could include an active choice of living space, preferred energy sources, a means of movement to obtain the necessities of life, and other volitional traits. These behaviors lead to a source of novelty.

Just as a bird needs two wings for balance in flight, the internal intention to interact, and an environment with which to do so, are both needed to explain the evolutionary process. For example, the separation of humans from other animals, including apes, included intentional activities such as standing up to allow the use of the hands. This alternate positioning of the upper body, in turn, allowed for activities that created demand for many internal changes such as varying the use of lung and vocal functions and the development of a larger brain. For example, detailed manipulation of the environment with the hands created much greater demand for tactile processing power that we think contributed to modern human brain development. Smaller and more specific changes might include such behaviors as wearing clothing, which likely induced bodily hair production to decrease over time because of reduced demand for it. Cooked foods induced digestive systems to change, as well as leading to the reduction of unneeded teeth and jawbone mass.

FIVE CHARACTERISTICS OF GENETIC PROCESSES

From a broad and long-term perspective, we think genetic processes have the following five characteristics:

Genes represent an organism's recorded history, developed and organized over time as an operating menu. The organism can, with time and need, add new materials and update old materials. This recording is based on intentional interaction with the organism's environment, and genome mutations can be based on its needs.

1. When new information is added to the genetic code, it tends to update but not erase previous information. Even if the gene structure or expressed function changes, the information will usually be passed on indefinitely.
2. All previously stored genetic material may be re-opened and used if the need arises and conditions allow. Such functional "reawakening" happens in the same order the functions were "turned off."
3. When unneeded functions are dismantled, the process will be the reverse of its development process, so that the first that appears will also be the last to disappear, and the last to appear will be the first to disappear.
4. We think there are three scenarios of genetic modification:
   A. Random occurrences: could happen to any organism.

B. Active modification or additions: happens in organisms based on need. We firmly believe that DNA is a tool that life has built during evolution and continues to alter when needed and possible. This accelerates the effect of natural selection and explains periods of very fast evolution.

C. Not a true mutation or novelty but an expression of older genetic information.

The points of discussion in this text are not entirely based on researched data, but are more from theory and extrapolation of the overall idea. We hope that the ideas contributed here add to and generate further evolutionary discussion and dialogue with those who have similar findings and ideas.

Part 1 of the book is meant to put forward a very simple and general framework for a new broader evolutionary theory. Part 2 uses the line of thinking that the intended evolution framework would lead to on various topics. We have not included any specific applications but, rather, ideas for further inquiry.

PART I

UNDERSTANDING
INTENDED
EVOLUTION

CHAPTER I

THE THEORY OF INTENDED EVOLUTION

—

A VARIETY OF EVIDENCE AND OBSERVATIONS HAS LED TO THE conclusion that the complex life-forms we see today, including humans, evolved from simpler forms over time. Darwin's evolutionary theory, natural selection, basically states that in a population of organisms within a particular environment, those that are most fit will live longer, produce more offspring and, over time, flourish and dominate the population. In this way, those best able to survive in their environment are "selected" by nature.

Over time, natural selection has been updated, with a number of variations, to reflect new scientific findings, including those from the field of genetics. The genetic material of an organism can be thought of as the basis of selection, and, in one prominent version, random changes in the genome (mutations) are said to create the variety of individuals from which nature selects survivors. In this model, random mutation and the environment (the selector) are the drivers of our evolution.

But, according to the theory of intended evolution, it is

an organism's internal intention to perceive, organize, save, and intentionally act on information from the environment that "drives" that organism to adjust and change internally. Before going further, we would like to say here that we are using the term *theory* in the general, and not scientific, sense in our discussions.

We believe evolution is an active, two-sided process through which life can intentionally change internally, and is therefore "shaped" by the environment, and not only passively selected by it. Therefore, an organism's intentional activities are the context for life's internal changes, such as manipulation of physiological processes including alteration of genetic material and evolutionary changes in form and function.

J. B. Lamarck, a contemporary of Darwin whose ideas competed with Darwin's, also proposed that an inner *life force* existed and that changes were based on what organisms did or needed to do. We propose that both concepts have merit and that, without an internal force, or intentional drive, activity that has been described as differentiating life from nonlife would be unlikely to occur.

Intended evolution's proposition of an internal, intentional force clearly parallels certain aspects of historical ideas such as vitalism, or the existence of a life force. A full discussion of this concept or of Darwin's and Lamarck's work is beyond the scope of this book; however, intended evolution is not meant to replace these important ideas but, rather, to pull them together and to elaborate on and recontextualize them.

Furthermore, understanding the importance and influence of intention, intelligence, and what we describe as life's basic process, the *information cycle*, may lead to potential insights into various areas of the biological sciences.

THE FORCE OF INTENTION AS AN ACTION POTENTIAL

By *intention*, we mean the potential to determine action (e.g., change or movement) on one's own initiative rather than only under the influence of some outside force. Living creatures can determine their actions, whereas nonliving objects cannot do so. For instance, a microbe can react to stimuli, but a rock can only be acted on by gravity, friction, or other natural forces. The term *intention* implies perception, intelligence, and potential action and is also interrelated to each of those factors. Whether the word *intention*, as opposed to some other descriptor, is the best fit may be an open question because these definitions tend to be circular. Generally, *intention* seems the most appropriate term to describe many of the characteristics normally used to differentiate life from nonliving things.

Historically, the term *life force* has been loosely defined as the vital or creative force in all organisms that describes their consistent behavior pattern, which differs from inanimate objects and is responsible for growth and evolution. Intended evolution's *intentional force* is an "action potential" toward a particular state or direction. When it is associated with what we think of as observable life, it is not unlike the relationship of the force of gravity with mass. The force of gravity itself cannot be seen, but the effects of its field can be observed as an object falling to the ground. Similarly, although we cannot observe the intentional force directly, we can observe the effects of the force in living things.

Therefore, we suggest that there is an intrinsic force of nature that is entangled within—or an emergent attribute of—the currently described natural forces that allows the emergence of life. This force has the attribute of being intentional in nature when combined with other preconditions of life and gives living things the properties that differentiate them from inanimate things. Speaking generally, we believe this is what gives living things the attribute of organizing or integrating themselves versus the tendency toward disorder or entropy.

We know that life cannot survive, nor could it have even come into existence, without the availability of certain building blocks, or prerequisites, such as water and carbon. We can think of these necessities as detailed levels or patterns of organization such as an atom of carbon or a molecule of water or, in a larger, combined way, like lakes or rivers. The most basic patterns of information form higher and higher ordered patterns, such as those described by the global water or mineral cycles or other ecosystem processes. Life on Earth somehow originally emerged amid these systems when conditions were right; it then expanded and evolved.

We believe the intentional force, when combined with other preconditions to life, provides a potential for action, or in our terms, it provides an *action potential*. This creates life's consistent pattern of behavior, or its ability to choose to change as distinct from *nonlife*, which changes not by choice but only when it is acted on by the other universal forces. Put simply, living things can affect their own futures.

THE RELATIONSHIP OF BASIC INTENTION WITH OTHER NATURAL FORCES

This does not imply that life operates contrary to natural laws, such as physics or chemistry. In fact, the opposite is implied: Since intelligent life emerged entangled within these forces, it is part of them, works within them, or, we could say, knows their rules. Therefore, like a fish intrinsically "knows" or "uses" the properties of water the best it can, emergent or simple life learned to use the basic forces for its own benefit when possible.

Life can choose to initiate an interaction with them and is able to affect possibilities at points where enough flexibility exists. By this, we mean, for example, that although life cannot necessarily cause or reverse chemical reactions, it can affect certain future chemical states if enough flexibility exists and it has enough energy at its disposal. An example is that when a system is close to a phase transition, it may be possible for life to choose the system's direction. Flexible situations are therefore very useful for life, both internally and in interactions with its environment.

Life, which we think of as anything that can intentionally interact with its environment to affect its own future, can choose from variable outcomes given access to enough energy for a given circumstance. Energy is the currency of change, so to speak, and any intentional change depends on it. This includes an organism simply retaining its internal structural equilibrium against external environmental forces. According to this viewpoint, it is the organism that constantly guides its own internal change on the basis of signals or information that it perceives from the environment. Furthermore, in some cases, the organism can also make changes in the environment to fit itself.

From this perspective, life is not separate from the rules of the universal forces that make it up. Instead, it is part of the fabric of these rules that are described by science and that, we could say, follows them but also uses them for its own benefit.

Potential for action is consistent with living things' observed ability to interact with their environment as well as other commonly noted attributes of life. We say action *potential* because this force, like gravity, is ubiquitous and may manifest only as preconditions are met. Therefore, a given state may have potential for action that is not activated until one or more preconditions are met.

For example, it can be observed that an organism may not react immediately to a given stimulus until other factors also come into play. Furthermore, in higher animals, an action potential may be a controlled intention that may not be acted on or only partly acted on. For example, a deer noticing a predator will let its flight potential build or dissipate while monitoring the situation. Perhaps potential turns to action only if the predator shows certain signs of approaching. Intentionality, therefore, is active in nature, and implies a potential for future change of state or direction. Looked at in this way, intentionality also implies predictions or projections based on accumulated information, by which we mean an estimate or forecast of a future situation or state. This concept will be clarified further in the next few chapters.

A PERSPECTIVE ON INFORMATION

As we will explain here and additionally in chapter 3, by *information* we mean what is conveyed or represented by a particular arrangement or sequence of conditions or things. With respect to living things, the sequence entails the *experience* of variable internal states over time, which change based on information patterns from the environment. Representations of these changes are then saved, sorted, and filed for future use. Therefore, information used by a life-form consists of saved representations of its experiences of environmental information patterns, which it later uses to represent the external environment. *Perception* is the internal comparison, or interface, of information from the environment with the saved representations of previous experiences.

As an organism accumulates or internalizes information through experience, it can project future outcomes on the basis of these past experiences. Action potentials can build or lessen on the basis of each new bit of information being used to update projections about the future. Before turning potential into action, an organism chooses between possible projections that match outcomes deemed to satisfy the current expression of its basic intentional drive (what it is intending at that time). Therefore, one of life's basic functions is to accumulate internal representations of useful information about the environment. These, and the ensuing statements, will be elaborated on in the next two chapters on information and a discussion of information processing.

A PERSPECTIVE ON KNOWLEDGE AND INTELLIGENCE

Organisms perceive information from their environment by applying previously acquired information or knowledge when deciding what to do, thereby exhibiting intelligence. We call *knowledge* the information gained by experience, and *intelligence* is the ability to use knowledge in the analysis of current perceptions. This analysis entails the internal arrangement of saved representations to form timed potential events, such that a projected outcome is available to be attempted. In other words, organisms formulate action potentials based on their previous knowledge or experience.

A basic activity of life is to use its knowledge base to make intelligent projections about future outcomes. These projections are then used to make choices about potential actions. Furthermore, the outcomes of those actions are perceived as experiences and are saved as new or updated knowledge. The circular nature of this process will be addressed in the next chapter.

. . .

The currently accepted versions of natural selection differ from the theory of intended evolution in that the former are focused on the external selection of random changes, whereas intended evolution proposes that internal changes can also be intentional in nature and based on intelligent interactions with the environment. Natural selection then acts on both these types of internal change. In a way, Darwin's ideas also

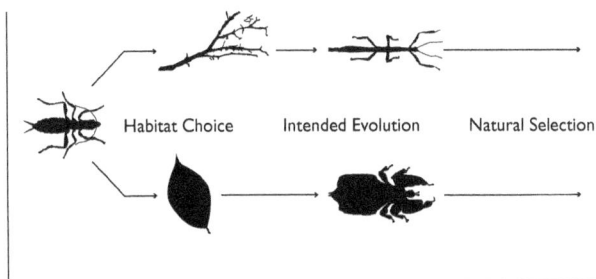

FIGURE I—NATURAL SELECTION DEPENDS ON THE INTENTION OF
AN ORGANISM
*This is meant to show that individual organisms make choices (e.g., its habitat). The
results of those choices (e.g., living in that habitat) act as input to natural selection
(e.g., fit individuals survive in that habitat), which is evolution.*

implied an internal component, in that he described natural selection in the context of "a struggle for life"; *struggle*
implies will and intentionality. We are putting forward that
the intended evolution framework is the missing piece that
Darwin alluded to at the conclusion of his introduction to *On
the Origin of Species.*

Finally, intended evolution theorizes that the role of the
intentional force in evolution increases in importance with
complexity. In advanced life-forms, the impact of their intentional actions can become a significant factor in their own
evolution. It is possible that a new phase will begin—or has
already begun—in which our internal intention and drive
become the main components of modern human evolution.

CHAPTER 2

THE INFORMATION CYCLE

—

THE THEORY OF INTENDED EVOLUTION SUGGESTS THAT THE processes attributable to life, including evolution, are best described as being intentional in nature. In this chapter, we propose that living things have a continuous cycle of intentional activity for which an action potential is the underlying driving force. The complex nature of the workings of intelligent processes makes discussion on the topic challenging. However, the purpose here is not an exhaustive discussion about the workings and no special or novel knowledge about intelligent processes is implied. Rather, we hope to simply formalize a process for the purposes of discussion.

When we say *continuous cycle of intentional activity*, we use the term *cycle* for illustrative purposes only; there is no implied limitation to flow characteristics: no beginning or end point. Rather, we hope that discussing the upcoming topics in terms of an action potential flowing through a cycle is helpful.

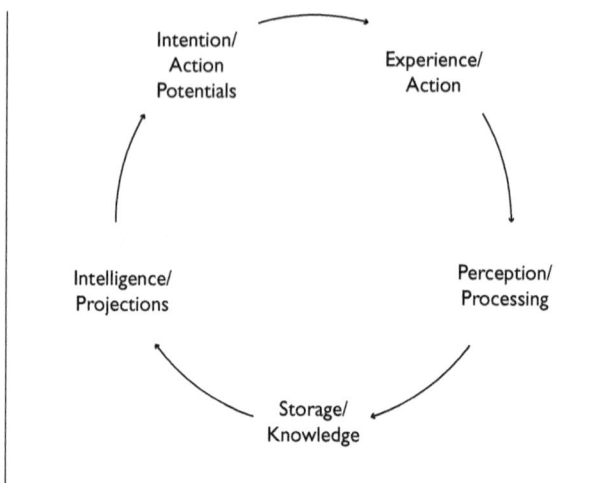

FIGURE 2—THE INFORMATION CYCLE
A cycle of input (experience), storage (memory), analysis (projections), and output (action).

THE ACCUMULATION AND ORGANIZATION OF INFORMATION

Any life-form, even a single-celled organism, experiences its environment by gaining information through internal change, or perception. When we say accumulation of information, it means internal change that leads to a subjective recognition of patterns. Life intelligently organizes the available information, which we could call knowledge, in order to project and effect possible outcomes of future events. For a given situation, available projections are evaluated, and a choice of action can be made, and then carried out. This in turn is a further perceived experience or interaction used to update previous knowledge, which continues the cycle.

We simply propose that knowledge is gained when experience of incoming information (changes in internal states) is repeated again and again, forming recognizable patterns. Representations of these patterns are then somehow stored, depending on how they affect the organism. For example, concepts like the relevance, repeatability, reliability, usefulness, and relationship to other (previous) knowledge are probably utilized. The point we are making has more to do with our idea of the process being intentional and intelligent than any rule set or mechanism of storage. We might also call this *memory*; we chose *knowledge* because this term implies purposely organized or sorted information as opposed to a random recollection of past data. How life internally organizes relevant information (e.g., the DNA molecule) is obviously complex and little understood.

Finally, we will often use the term *information cycle* as a way to discuss any organism's ongoing interaction with its environment going forward.

ENERGY AS THE CURRENCY OF CHANGE

Before moving on, we want to briefly mention the energy used by life as somewhat distinct from other patterns. We view what is commonly referred to as *energy* as the currency of change and believe that life uses it purposely in its processes. Having access to reliable energy sources is obviously a big factor in how the environment shapes evolution. It is not surprising that life views information about energy sources and usage as important since it is a basic need to bring about desired changes or even just to maintain an organism's current form in a changing environment.

A SIMPLE INFORMATION CYCLE EXAMPLE

Using a simple microorganism to discuss the information cycle, let us assume that information in its environment is a recognized energy source. To an amoeba, for example, a paramecium is a recognized energy source that is regularly consumed. When we say *recognized energy source*, we mean it has been experienced previously and a representation of it has been stored as knowledge that can be used in projecting future outcomes. Therefore, according to our model, this information (*paramecium*) has already been repeatedly perceived and evaluated and is projected to be relevant as food when perceived in the future. Furthermore, repeated choices have been made to take action (e.g., to move toward the food source and consume it), and representations of those experiences stored as knowledge. Repeated or important experiences are stored and updated, depending on factors such as relevance, reliability, and so on. When a food is experienced again and again, projections about potential outcomes become more reliable, and action potential can turn into action more quickly. The action stage, too, is perceived and saved as knowledge, starting the cycle over. New, updated knowledge then becomes the basis on which to make further projections about information perceived in the environment. As stored knowledge gets more reliable, the projections based on it become more reliable, which leads to more rapid and successful choices and actions. This, in turn, frees up perceptive ability for other activities and, therefore, increases the efficiency, scope, or, we could say, the ability of the cycle (i.e., through increased perceptive abilities).

THE SUBJECTIVITY
OF EXPERIENCE

Before an action is initiated, intelligence is used for evaluation, formulating projections, and building action potentials. This evaluative part of the cycle can vary, depending on need. If the same information is acted on in the same way over and over, less intelligent evaluation is needed, and perceptions can be acted on more quickly. Therefore, reliable information and knowledge are very important and can save a lot of time and energy. However, more or longer evaluation of choices can also occur, and action potentials may or may not result in action. Furthermore, information that has been rarely perceived or that is not contextually relevant may result in little or no action potential unless it is perceived again and some context is filled in. We call this *raw* information; it lacks the context to make it relevant or reliable at the present time. If raw information is repeatedly perceived and some context is gained with each iteration, it can be evaluated more fully, can be used for projections leading to experienced action, and can become information saved as knowledge. In this way, a knowledge library is built and updated through repeated information cycle activity.

Interestingly, it is the previous experience or knowledge that gives perception context and affects what is recognized and the way it is recognized. Therefore, perception is based on previous experience and knowledge and is itself subjective. Because perception is subjective, all experience, as well as saved knowledge, is also subjective, as is the entire information cycle process. Of course, projections are therefore also based on subjective information.

When raw, unrecognized information is recognized over time, the process is based on previous knowledge, but also updates it, or gives new context or meaning to that knowledge. Perceptive ability and characteristics are therefore also updated to reflect the new version of the "knowledge library." We could say that the information cycle also grows with and in the direction of knowledge updates, as long as the knowledge base is protected and the information cycle continues. Because perception and projections are based on previous knowledge and experience, it is crucial that the knowledge library is protected, updated, and reliable.

NEW KNOWLEDGE AND THE RECONTEXTUALIZATION OF OLD KNOWLEDGE

With the information cycle framework as it pertains to evolution, we observe a repeated layering of intelligent and intentional activity. We say *layered* because new knowledge implies integration with—and recontextualization of—previous knowledge. Because the information cycle, including the accumulation of new knowledge, operates on the basis of old knowledge, change is limited and manifests differently for each organism.

A CONTRAST OF UNIVERSAL AND SUBJECTIVE ENVIRONMENTS

When we talk about the external environment that an organism perceives and interacts with, we are talking about a subjective environment or what we call its effective environment. An organism's *effective environment* is the scope of the external environment that the life-form can perceive or interact with. The information cycle is therefore a subjective process, because each organism will perceive a different effective environment; that is, it perceives a version different from the actual or *universal environment,* which can be thought of as the totality of actual reality, without individual subjectivity.

Different organisms' effective environments occupy the same universal environment, but each organism perceives only a portion of the universal environment. Furthermore, any piece of recognized information or factor in the universal environment is also recognized subjectively for the same reasons. Effective environments, therefore, are an incomplete picture of the universal environment but grow and become more complete as knowledge increases, such that the information cycle approaches but can never achieve a perfect

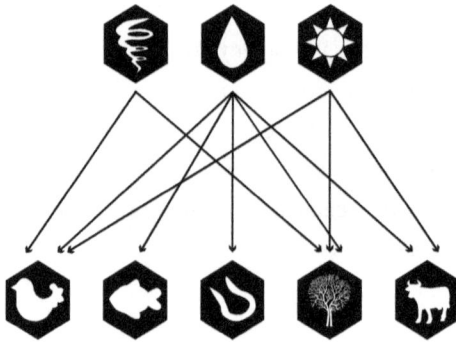

FIGURE 3—SUBJECTIVE ENVIRONMENTAL FACTORS
In a shared environment, each individual life-form's subjective environment is different.

representation of the universal environment. As the effective environment grows more similar to the universal environment, the organism's projective ability increases, reflecting its more complete and reliable knowledge library.

FACTORS WITHIN AN EFFECTIVE ENVIRONMENT

We can think of recognized things in the effective environment as *effective environmental factors*, or a given piece of information in the environment that a life-form can perceive and interact with. For example, a piece of wood that is recognized as food for a microorganism may be recognized as shelter for a larger life-form or as a tool for yet another. In this case, the piece of wood exists in what we are calling the

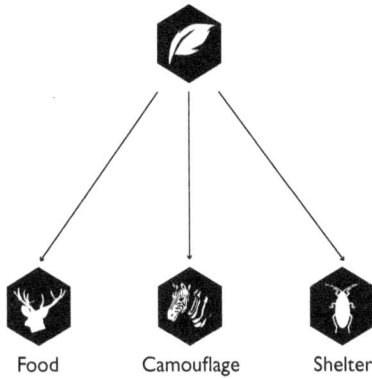

FIGURE 4—SUBJECTIVE ENVIRONMENTAL FACTORS
A single item can be used for many purposes, depending on the intention of the user.

universal context, but each life-form views it subjectively in its own context (of course, that description—*wood*—is also a subjective, human idea and not actually universal). Organisms can cooperate and share effective environmental factors (e.g., wolves in a cave), can compete for effective environmental factors (fighting over food), can use the same effective environmental factor differently (the wood example, above), or some combination of these such as sharing an effective environmental factor despite having different effective environments and so on. Therefore, the sum of a life-form's effective environmental factors is its effective environment.

In this way, the environment can be divided up into pieces for the sake of discussion about an organism's perceptive ability, functions, and interactions. For example, certain wavelengths of light, which we perceive as colors, are effective environmental factors for some organisms but not for others (e.g., photosynthesis, varying levels of spectrum recognition). An example of an organism perceiving and using the same universal environmental factor differently is photosynthetic life. With the same universal environmental factor (sunlight), differing life-forms use different recognized information (pigments) to make use of a different part of the light spectrum, presumably on the basis of its availability in different organisms' effective environments. Light is a universal environmental factor that is a readily available and stable pattern of information available on Earth, so it is not surprising that most life uses it in some way.

THE VARYING SCOPE OF PERCEPTION

Simple life-forms have recognized relatively few effective environmental factors, and therefore have limited perceptive ability. In other words, they have a very limited effective environment. A lack of recognized information from the environment implies a relatively smaller information cycle, so there are few potential projections to chose from or actions to take. In such cases, the reliability of information is predominant; flexibility and intelligence are limited, as is the ability to use energy. This applies to the first evolved life-forms on Earth and also to simple life-forms that exist today. Because choices and actions are limited, intentional environmental interaction is also limited, and the traditional version of natural selection is the predominant factor in the evolution of these simple life-forms. However, we believe that because intelligent choices and actions also get selected with larger effective environments (organisms with greater perception and intelligence), they become more important selectable traits as more complex life-forms evolve.

Single-celled organisms, for example, have a smaller scope of intelligent ability and can therefore "intend" to do very little planning to avoid environmental challenges. The wording is difficult here, because we are not saying these organisms aren't intelligent within their limitations. Rather, because of limitations in scope of awareness their survival is generally challenged directly, and they have to resort to drastic attempts to change core internal functions and structures in order to survive.

Their evolution is usually more likely to be due to selection by environmental factors that they can do little about, because those factors would appear random to them. A mammal, however, can perceive and adjust to large-scale information patterns, such as weather and seasons (e.g., as a signal to migrate), can adapt internally to new food sources or other challenges, and can learn to avoid new predators. In more complex life, perception and intelligence are important traits that selection acts on. An example of intelligent adjustment by more complex life is the elaborate food-seeking behavior that we observe: digging, pecking, hunting, and even farming. These activities can be seen as being driven by the same basic intention as a microorganism's movement toward food and yet also demonstrate that differing abilities can arise to fulfill the same basic intention, depending on the organism's evolutionary history.

Perception and intelligence imply learned information or patterns that are, by definition, nonrandom. We discuss information more fully in the next chapter. Because the information in the environment that life interacts with is nonrandom, the evolutionary process will generate predictable patterns that can be used to develop strategies. We could also state this by saying that the process is based on "rules," or the characteristics of the forces of physics or chemistry. We want to reiterate that in an ever-changing environment, life must carry out the information cycle and evolve, simply to survive. This back-and-forth interaction described by intended evolution might be thought of as a game between an organism and its effective environment. This game has an intelligent player (the organism) who follows the rules (i.e., natural laws) perceived in its effective environment and acts accordingly to survive and prosper.

A CARD GAME AS AN ILLUSTRATIVE EXAMPLE

Many games have quite simple rules from which complexity can result. Life's evolution and its continuous information cycle—reading signals from its external environment based on its internal library—can be compared with playing a card game. Of course, the first step in any action is intentional drive, and, in this example, it is the intention to play. Players (organisms) use their perception and intelligence to interact with other players and the cards (effective environmental factors), according to the rules (the effective environment). As play proceeds, the players use the cards in their hand (collected information) as context for what is going on as play progresses. New information (cards revealed during play) update the old information (the cards in the player's hand) during play and is used to build strategies (action potentials) based on game knowledge (knowledge library) as well as to build a better hand for the next round. Success depends on how well they know and learn the game (intelligence). Continuing to play results in the players getting better at the game as they recognize new patterns of play and save them as knowledge.

If a new card is introduced, that new effective environmental factor is recognized and added to the knowledge base in the context of previous knowledge. This novelty may also alter the understanding of existing cards or combinations, which become better understood. With each round, knowledge is saved and updated, and the new projections about future outcomes are more accurate, which results in better play. Random chance or poor strategic play can lead to being eliminated from the game. The latter could be related to a shortfall in intelligence, a lack of focused intention to play, or other factors.

Under the right circumstances, a player might actively look for a card he needs to complete an intended situation or might wait for a situation to mature before executing a certain strategy (hold action potentials while awaiting new information). He or she may also learn new combinations or strategies through experimentation (novelty) or from other players and intentionally seek out new knowledge. We also often see card players put certain "no-brainer" plays on automatic, wherein little thought is given to well-understood or often-repeated plays. This automatic behavior is analogous to trained responses to the same stimulus, such as reflexes or instinct. Because it is based on repeated action, there is little need for processing. This allows players to use their intelligence to plan ahead to evaluate future possibilities, increasing their perceptive abilities dedicated to learning the game.

To say that the game is determined by the rules (i.e., it is deterministic) makes sense (with respect to the universal forces). To say that drawing new cards are important to the player's chances (the random chance) is also correct. But the player has to intend to play (i.e., to have an internal drive) for any of this to happen and becomes better at the process in doing so. It is true that sometimes, a good player may lose or a poor player may win, which means that chance can play a significant role, depending on the rules of the game. But in the long run, how the player plays the game (intelligence) and how well he or she learns more about the game (the environment) are the most important factors in complex games. Using intelligence, an organism learns how a game works over time, and those who keep playing (i.e., whose genetic lineage continues) become better players (i.e., they evolve).

CHAPTER 3

MEMORY, PERCEPTION, AND PROJECTION

—

THE INFORMATION CYCLE—THE ONGOING PERCEPTION OF an organism's external environment and the internalization of relevant information—is a basic process of life according to intentional evolution. Let us now take a closer look at what is meant by *perception of information.* Any perception implies an interface between an organism's internal attributes—including knowledge—and incoming information from external sources. In other words, the perception of environmental information implies an interface and comparison to the self, which is the context of the information. Even a rock reacts to the environment on the basis of its internal attributes, or, we could say, historical context. Different types of rocks react differently, depending on their internal structure or attributes. Furthermore, a given rock reacts differently today than it would have in the past because of internal changes caused by previous interactions. We can see that we could use the same or parallel terms for the rock as we did for organisms here and in the previous chapter. For illustrative purposes, we could say that a rock interfaces with or "experiences" only "recognized" information, meaning that there will only be interaction based on its unique internal structure.

We could also say that its structure consists of a "memory," in that it is a reflection of the past and past interactions, but we do not call it *knowledge*, because it has not been organized intelligently as was described earlier for living things. Therefore, rocks have internalized representations of past events, such as past environmental conditions, although to the extent we would call this "information," it is useful to us but not the rock. The events themselves are not present; only their effects remain. For example, energy passing through a rock may change the rock's structure, but the totality of the energy pattern is not in the rock; it has perhaps gone right through, with slight changes to the energy, to the rock, or to both.

Similarly, internal changes happen to living things, depending on their internal attributes, such as structure, memory, knowledge, and intelligence. As we mentioned previously, life can intentionally sort, save, manipulate, and act on information, which a rock cannot do.

HISTORICAL INFORMATION CONTAINED WITHIN STRUCTURES

All of life's internal structures and processes are also a form of subjective memory, or a representation of its evolution. Furthermore, life's perceived information must also be based on its history. This means that all future information interfaces of any life-form will depend on its history, including knowledge.

For example, while reading, we compare symbols in a book to our internal knowledge. If the relevant language has not been learned, the book cannot be read. The reader must first expose him- or herself to other information (e.g., the rules of a language) and update or recontextualize his or her knowledge library in order to gain context to understand the new pattern (the language).

Evolution works in the same way but over longer time frames. An organism's structures and functions are based on perception of the effective environment, including the building blocks of life that make it up. For example, if there is no visible light, no use for it can be derived, and a life-form will not build eyes or light receptors. If light becomes an effective environmental factor for a given life-form (e.g., it relocates to an area in which light exists), it may then learn to recognize and start to use light, creating the need to build structures to do so. We also see that life will dismantle or change structures over time if their previous interface is no longer available. For instance, fish that move into caves may lose their need for vision and then the physical structure of their eyes will tend to change over time to a more functional form.

THE PERCEIVED BASIC CHARACTERISTICS OF INFORMATION

Everything that life can sense comes from information patterns, which can be compared to waves. Waves have characteristics such as a frequency, which implies a time factor, as we have noted. The concept of time, then, is simply notice-able changes in factors, such as the changes that we call *days* and *nights*, but also moment to moment changes.

Perception occurs when our senses pick up incoming energy in a recognizable form or spectrum. For example, there is much electromagnetic energy that we do not see and many sound waves that we cannot hear. The patterns that we sense are what was labeled as *recognized information* in the previous chapter, whereas what we cannot sense, we labeled the *raw* or an *unrecognizable* portion of the universal environment.

The main activity of life is recognizing useful patterns and saving internal representations of them for future use. As opposed to the rock in the example above, life-forms develop their own sensory systems over time to perceive, save, and use representations of information patterns that have been recognized to be useful. These living systems intentionally develop as information in the environment is recognized as beneficial and those life-forms and their internal systems are induced to take advantage of it.

Life saves internal representations of useful segments of information as memory and knowledge. In other words, life can choose what part of a pattern it saves and how to save it on the basis of its recognized usefulness. For example, there is no reason to save a representation of an hour's worth of a given sound if a three-second segment gives you something of equal value. This is why one experienced segment of a pattern is used to extrapolate other iterations of the pattern. An organ-ism's knowledge and memory are subjectively saved represen-tations of segments of patterns that are used to project future outcomes and to create its action potentials. Therefore, when an organism draws on its knowledge of the past, that process also involves a projection and is subjective. This means that knowledge and memory, like ongoing perception of information, is subjective. Experience and its timing are therefore a subjectively "filled in" process based on a comparison of subjectively relevant past events (which were subjectively saved) and information from the environment, which is, itself, recognized on the basis of previous experience.

This seems clear, because a given memory was subjectively experienced and saved, but when new information is recognized and perceptive ability is increased, it also includes and updates perceptions of the past. In other words, an organism's understanding of its own knowledge and memory changes, or is recontextualized, with the recognition of new information.

Complex organisms such as humans can notice fluctuations or inconsistency in patterns of longer duration or that are more complex, such as comparing the daily fluctuation of light or weather to seasonal ones. Patterns such as weather, seasons, or even day and night exhibit varying degrees of change, and, over time, they are not completely consistent; thus, life must constantly adapt to ever-changing environments in order to survive.

We should briefly mention that the use of the terms *pattern* and *information* is meant to differentiate from the idea of randomness, which, by definition, is not recognizable to intelligent life: It is a lack of pattern or order. Anything nonrandom has a temporal aspect to it, because a sequence or arrangement implies that one event occurs after another in time.

THE CHARACTERISTICS OF RELIABILITY REGARDING INFORMATION

This means that when it comes to reliability, the more subjectively random-looking or less recognizable a pattern is, the less useful that pattern will be. This could be due to exposure to the pattern, which can make it subjectively more reliable over time as it is more fully recognized. For example, a pattern only seen periodically or a pattern with, say, three different oscillations will be less reliable than one continuous pattern—until the new pattern is fully recognized. Generally, the more regular and reliable it is, the more useful it will be, whereas irregularities lead to less usefulness. This is because the reliability of a pattern increases the accuracy of projections derived from it.

As an example, a spoken language in a given area is a recognized verbal pattern, which makes it a useful system and a very valuable pattern. But, as we have noted, the less reliable the pattern is, the less useful or efficient it is. So, uncertainty about the meaning of words requires more time and intelligence during interactions. Furthermore, if the pattern is less universal (e.g., not everyone in the area speaks the same language), it is also less useful, because it is much more difficult to make projections about what others are going to do if they cannot be understood.

When making projections about a given effective environmental factor, accuracy will decline with distance in space and time from the projector. In other words, a saved representation of any pattern will yield more accurate estimates of future outcomes the closer it is in time and space. For example, projections about future food availability (like all patterns) will be less accurate the farther in the future we try to plan for.

Furthermore, as the scope of perception of an organism increases to include more time and space, a given pattern will tend to change in meaning and will eventually become part of a larger pattern or context. Conversely, theoretically speaking, as scope decreases, what were once viewed as patterns can break down and appear random. A larger scope is needed to understand seasonal patterns, the recognition of which recontextualizes the meaning of day-to-day weather.

This is not to say that effective environmental factors that stretch over longer time frames are less reliable or less useful than shorter ones. Seasons are temporally longer "waves," or effective environmental factors than a day's weather but are more useful for certain purposes. However, given the effective environmental factor of seasons for a given organism, it is generally more reliable and useful in projecting plans for the next few years than for 1,000 years from now.

Consider a rock that a lizard lives on. It is a very stable and effective environmental factor for the lizard; the rock's properties are about the same every time the lizard sees it, possibly for the entire duration of the lizard's lifetime. It knows every inch of the rock and all the hiding spots, which makes the rock a stable and reliable effective environmental factor, assuming that the lizard has repeated experience using it. Like the seasons, this effective environmental factor is very stable over time. However, the rock is recognizable to the lizard in much less time than are the seasons, because the information obtained by interacting with the rock is very stable and fixed in the short term. On the other hand, the seasons must be experienced over a longer stretch of time in order to be recognized as patterns.

THE SUBJECTIVE UTILIZATION OF ENVIRONMENTAL FACTORS

Although stable information and patterns are more reliable, they are not always more recognizable. The seasons are very reliable patterns, but they are not recognizable (i.e., not an effective environmental factor) for many organisms. If an organism's life span is only a few weeks, seasons cannot be recognized as an effective environmental factor. More broadly, an organism's ability to recognize patterns is limited by its scope of perceptive ability. Furthermore, slight differences in pattern recognition could lead to similar environmental factors being used in many different ways, depending on the internal knowledge base (or context) of the organism that has recognized it. Technically, what we call the "universal pattern" of seasons is recognized as a different effective environmental factor, depending on which organism is recognizing and using it. We are assuming the term *seasons* is more universal in nature for most of life, but it is not implied that all animals understand what a season is in the way that humans do. Some animals have simply been observed to utilize certain cues to alert them to future conditions, and they react to these cues in their own ways. For example, birds migrate but bears hibernate in response to seasonal changes. Both have "recognized" seasons as a very reliable effective environmental factor in their own subjective context, and so each life-form uses or responds differently on the basis of its knowledge and intelligence.

When we say an organism "responds" to the seasons or other patterns, we are also including the internal changes and adjustments individual organisms have made over time that make them recognizable as, say, birds or bears on the basis of the choices they made in evolutionary history. At some point, these organisms recognized given environmental factors differently (i.e., they had a different context with which to understand a given pattern of information), which resulted in their evolutionary divergence in form and function.

We believe information about effective environmental factors is also recognized and processed according to its perceived importance, not just on the scale of life and death but also on a scale of ease and value of use. For example, a more reliable food source will be given priority by repeated interaction and information about it accordingly sorted, evaluated, and stored. Furthermore, this value may have many aspects, depending on the perceptive abilities of the organism. In general, as with any information, reliability and consistency are important, as is shown in the example above with respect to languages. An example that is relevant to most life would be sunlight, which is a widespread effective environmental factor because of its abundance, consistency, and usefulness as an energy source; it is therefore "recognized" and utilized by much of life.

Finally, other organisms (e.g., predators) can obviously also be very important effective environmental factors, because they might threaten or benefit the structure and knowledge directly. Instinctual responses and self-preservation are discussed in the later chapters.

LEVELS OF ENVIRONMENTAL FACTOR AVAILABILITY

In general, a changing environment means more available environmental interaction. Life will experience and recognize more effective environmental factors over time via the information cycle simply because of their availability. This means that the entire information cycle process increases in size, including increased knowledge and perception over time. In very stable environments, after the available effective environmental factors are recognized, little change will occur other than offshoots of previous change. The available raw information has been recognized as effective environmental factors, and little perceptual change takes place as life waits for new information.

Situations in which evolution seems not to have progressed for long periods of time make sense in the information cycle model. Lags in any part of the cycle—perception (e.g., no new information that is recognizable), intelligence (e.g., an inability to do anything with new patterns), or action (e.g., an inability of an organism to change itself)— will hold back change. Patterns that are subjectively valuable to an organism or groups of organisms can also lead to the intention of continued use or more elaborate use of a given beneficial pattern rather than diversification to other resources. The abundance of sunlight, for example, led to the proliferation of photosynthetic organisms.

Although we have said that increased perception allows the recognition of larger effective environmental factors, this is difficult wording, because it actually implies a greater or different understanding overall. There might be an understanding that a given effective environmental factor may be larger than was previously thought, had more uses, was part of another effective environmental factor, or was seen in a more detailed manner. For example, an organism may connect temperature changes with leaf color changes, which may increase its understanding of a concept closer to what we think of as seasons. Perceptual increases mean that previously disparate effective environmental factors are revealed to have a common context as the effective environment expands and encompasses more of the universal environment. As evolution "advances" in this way and organisms become more complex, the entire information cycle becomes larger: greater perceptive and projective ability, higher intellectual capacity, and more choice and action possibilities.

LOOKING AT INFORMATION AS REPRESENTATIONS OF THE PAST, PRESENT, AND FUTURE

Another way to look at the nature of information is its place in time as past, present, and future information using our subjective information cycle framework. Past information can be looked at as any form of knowledge library or memory, present information is what is perceived to be going on now, and future information is what is projected to happen in the future. All information referred to in the information cycle, including knowledge, is a subjective internal representation, based on perceptions. When something is perceived, we could say that an internal change has occurred from one state to another—in other words, perception is a comparison of internal states.

PAST INFORMATION

Past information (memory or knowledge) is subjective and not *true* in the universal sense—neither now nor when it was saved. Rather, it is a stored representation of a subjective experience. Furthermore, past experiences can be based on projections of the future that were deemed to be the best option at that time. Previous experiences were also based on what was past experience or knowledge at that time. Any experience at any time has to have a previous basis for understanding. Current experience, then, depends on—and yet also changes or recontextualizes—past experience and knowledge. In other words, memory and knowledge are changed by ensuing experiences that are used to update them, meaning they are constantly changing and being updated.

In a universal environmental sense, a past event really did happen; it doesn't change. However, the knowledge base that we are talking about is a subjective representation inside of the organism of a given event and is constantly being updated.

Therefore, new information changes the context of all other information but can be seen as *soft* (in flux) rather than *hard* (more stable). This is because older, "harder" information is evolutionarily more secure in that it has already been updated and changed many times. If an experience has been updated over long periods through repetition, it develops into the very important and more rigid systems that evolution has shown to no longer need much change, if any. When information is first recognized, its internal representation can be updated easily. Over time, it gets recontextualized over and over again through connections to more and more information. It is these outer, more flexible layers that get updated first, while the internal information and systems stay relatively stable. Therefore, similar to future information (projections), the farther away from the present time or location a projection is, the less clear it will be about the past, because details are lost with respect to current perception. Essentially, the older the memory is, the more it has been updated from the original context, and the original meaning of events is changed and becomes less clear.

FIGURE 5—THE CLARITY OF PROJECTIONS DECREASES WITH DISTANCE FROM THE PRESENT
A time line in which the center (the present) is the clearest, most visible, most well-lit information; the past and future both become less clear as distance increases from the present.

PRESENT INFORMATION

The present is the point at which experience, action, and perception all happen and at which the present effective environment is recorded in the knowledge base. Although the decision of whether to act is based on past information (knowledge), the present is the transition or choice point between the past and the future. Current choices and action "harden" the flexibility of projections and action potentials into memory and knowledge. As we said earlier, action potentials can build over time, and projections narrow until choices are made in the present moment. When action is taken in the present moment, it is also experienced as a perception and recorded, starting the cycle again. However, the notion of starting and ending is for descriptive purposes only; the cycle is continuous. We should mention that raw or relatively unrecognized pattern collection also happens in the present moment, as does the recognition of previously raw patterns.

As we have said, our perception of present information is based on previous memories or knowledge. For example, if we encounter someone saying something relevant, it could be used to project a future event, could be used to recontextualize our knowledge base, or both. But someone saying the same thing in a different language would be treated as unrecognizable information until repeated experiences lead to recognition over time. Therefore, the recognition or meaning of incoming information in the present moment is based on existing internal information—knowledge.

FUTURE INFORMATION

What we are talking about here are projections or estimates about the future. Projections about the future are made with the most up-to-date knowledge used as the basis for estimation. Therefore, the act of creating projections is based on stored knowledge with the addition of information from the current moment. Future projections range from relatively reliable projections about moments close to the present to unknown or unrecognizable situations in the unforeseeable future. In general, a given effective environmental factor is more reliable when it is close to the present time and space, and projections become less clear the farther one looks into the future. Projecting farther means that detail is lost; only outlines will remain, even when using the most reliable patterns or knowledge for projection.

A VIEW OF INFORMATION PHASES AS PHYSICAL STATES

Using a snowball as a loose analogy, the dark volume in figure 6 represents knowledge or the past, the surface area represents the present, and the outer bands represent the future.

Moving from the unknown future to the present can be looked at like water vapor condensing into clouds and then further into rain or snow. The information cycle turns raw or unrecognizable information (the water vapor), into the recognized effective environment (the cloud), which can then be recontextualized further (recognized more fully) as snow falling. Physical changes occur at each stage from recognition to action in the present moment, when action potentials turn to choice or action (condensed snow falling). An experience creates a kernel of knowledge—the center of a new snowball. As repeated cycles are saved to the knowledge library, the snowball becomes bigger, and the older knowledge becomes more internalized and solid.

Past information occupies a range from soft (recently saved) to hard (knowledge used and updated over long periods of time). Present information is the most current perception—the interface between knowledge and future projections. To some extent, experiences in the present moment were projected in the past and are now being realized in some form. They are used to update the knowledge base but, of course, are also subjectively based on it. As far as the future is concerned, the information closer to the surface is closer to the present time and is, therefore, more reliable

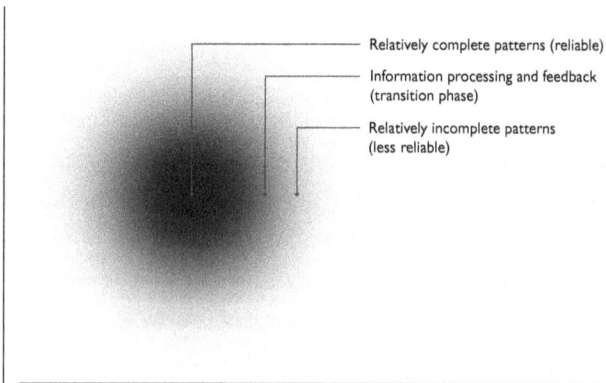

Relatively complete patterns (reliable)

Information processing and feedback (transition phase)

Relatively incomplete patterns (less reliable)

FIGURE 6—THE RELIABILITY OF INFORMATION AS A SNOWBALL
Older and harder at the center, newer and softer with distance from the center.

and certain. It is also, therefore, more likely to be in the effective environment, whereas information farther away is less reliable and less likely to be in the effective environment.

RAW INFORMATION PROCESSING

To follow this idea, let us look at the information cycle as it pertains to processing information, including raw (unrecognized) information into memory and knowledge (recognized and saved information). At first, a pattern may not match and make changes to current internal patterns (we sometimes refer to this as *knowledge*), but if it is experienced over longer periods of time or in conjunction with other information patterns, a pattern may at some point relate to and change current knowledge. At this time, some segment of knowledge (memory) is recontextualized (changed) by the new pattern and is recognized by the perceiver. At that time, an internal physical or energetic change has happened such that a previously unrecognized pattern is represented in the perceiver's memory. If it continues to have enough value and proves to be relevant, it will continue to be used to recontextualize deeper or older levels of memory and knowledge. We must stress here that the wording is difficult; we tend to use the term *pattern* in the universal sense, but it is, of course, recognized subjectively, so we are not saying that any universal information is ever truly recognized.

Valuable patterns perceived over many generations may become very hard and not so easily changed, becoming deep levels of memory, likely to be passed on to future generations. We tend to think that the current view of the chemical sequence part of DNA information is a generational time frame tool: purposely updated for the next generation and not to be easily modified by short-term changes in environmental information.

Harder, less flexible, information has been through more information cycles and connects to more parts of the knowledge library; it has been sorted, updated, filed, and used over time—just as the center of a snowball freezes harder as it grows. New information would tend to be more easily changed until it becomes more experienced. It would also tend to be used and influenced by action potential—but not necessarily action—until its context becomes more fully recognized.

Creating desired change in harder information, such as core physical structures (e.g., skeletal or organ structure) is more difficult and takes more time and energy. The very nature of such structures is that they are meant to be more difficult to change; they are the result of longer periods of reliable information, having condensed into a stable form that is deemed necessary for projected effective environments. Therefore, new conditions making old, reliable, hard information outmoded must persist before an alteration will be undertaken. For instance, if an organism that recognizes a new predator in its effective environment must make adjustments, the perception and intelligence portion of the information cycle process is more flexible, easier, and more readily available: the recognition of trouble and a strategy for escape using the current structure. Can the organism solve the problems with its current structure and by using its intelligence (e.g., to change avoidance behavior)? Persistent conditions may demand physical change, but this is much more difficult to achieve (chapter 7).

In summary, information is processed so that it can be made useful; it must be recognized, sorted, and filed (turned into knowledge) permitting projections about the future as well as the updating of previous knowledge. An actual information cycle process beyond our simple model would, no doubt, be very complex, and the development of some sort of protocol for knowledge manipulation and storage seems a worthy endeavor. We are not claiming any unique insights about mental processing but simply putting forward a simple model for discussion purposes, suggesting that such a process is integral to life and evolution.

INFORMATION PROCESSING ACROSS GENERATIONS

It is our opinion that projected changes may be recognized as needed, but rather than being acted on in a given generation, the representation can be passed down as an action potential to be acted on or added to by the next generation, when much more energy and flexibility is available—for example, during development. This method can be much more efficient than effecting physical change in the current generation, because changes are easier, involving more flexible versions of a given organism (see chapter 11).

Some physical changes (such as new structures) also take much more energy than others and are more efficiently built in the next generation. As an analogy, consider needed alterations to a building's interior based on a new demand that a company may face. An assessment is made, and, sometimes, if the changes are extensive or costly, it may make more sense to build a new structure rather than trying to alter the current one. Passing down action potentials would depend on such factors as the reliability, length, and importance of an information sequence, as well as its persistence over generations. Therefore, if there is a reversion to previous conditions, action potential will also begin to fade rather than build.

For example, in our view, annual patterns such as seasons could not be passed down until they have been recognized, which could potentially take more than one generation. Passing down seasonal information would allow such a long-term pattern to become a very reliable piece of knowledge. Similarly, a memory filed so as to be passed down to offspring would be deemed of long-term importance and would probably have been experienced over many generations and deemed important enough to be held in memory.

Clearly, in our model, change in all memory or knowledge, including DNA, involves an intelligent process. Therefore, according to intended evolution, DNA is a tool built by an organism in order to preserve information about itself with which to represent information about the environment and respond to its challenges, including the plans for future generations and their challenges. Although some random mutation makes sense, especially in simple life-forms, according to intended evolution, changes in DNA can be done intentionally when the need is great enough and conditions allow.

With respect to intentional mutations and changes over generations, we do not claim to know the mechanism for or the hierarchy of intentional mutation or random mutations. We believe that as life becomes more complex and exhibits increased internal control, intended mutations accelerate along with the ability to shuffle, file, protect, and correct the genome. Furthermore, this issue should be looked at with respect to internal systems' being intelligent and making subjective decisions about restructuring (see chapter 5).

INFORMATIONAL EVOLUTION

Natural selection tends to work primarily on past and present information (knowledge) in which the current knowledge library is selected by the current universal environment. Intended evolution pertains more to the future (projection) phase and the current intelligence phase, in which projections are created and evaluated. We believe that intended evolution is like the input to natural selection, because not only are intelligent behaviors selected, but the knowledge library and the structure of an organism that is selected was also intelligently constructed.

We think of modern synthesis tending toward one structure winning out over another. Intended evolution also sees the structures as being developed as an intelligent solution to an environmental demand and the more intelligent processes also winning out, especially in more complex life-forms.

We do believe that creating desired evolutionary physical change would be relatively difficult through intention alone, but need-based challenges are a major catalyst of change by intention and actions. Actions reinforce importance and create different type of memory and knowledge than intention alone. Historical, need-based challenges might include ocean-habitat mammals that require improved swimming ability or changes to the breathing apparatus in order to thrive or survive, or the giraffe needing to reach food that is far off the ground, as Lamarck noted. These changes are also more energetically expensive, take longer, and are multigenerational in scope, much like learning to utilize the changing of the seasons.

CHAPTER 4

CORE STABILITY AND DEGREES OF FLEXIBILITY

—

HAVING TALKED ABOUT THE PERCEPTION AND INTELLIGENCE (information processing) portions of the information cycle, we now want to address the execution phase. Because the intellectual phases lead to and experience the action phase, perception and intelligence are necessarily also involved, but we will be discussing action and the more physical attributes in this chapter.

As we have said, all of life follows the natural laws, such as those of physics and chemistry. But we also believe that although it operates according to those laws, life can also intentionally take advantage of them, including altering possible outcomes; life can make choices when there is enough energy and inherent flexibility. For example, we are saying that cells intelligently manipulate and regulate their metabolism to the extent that the available energy allows. They can make choices about the direction of a process or system that is flexible enough if enough energy can be applied to activate it. Life is as efficient as possible; it uses physical and chemical processes primarily to take advantage of the physical laws instead of fighting against them. The basic natural laws are therefore used as tools whenever possible.

First, we want to briefly discuss the terms *stability* and *flexibility* as they relate to our discussion going forward. By *stable*, we mean relatively central characteristics, structures, or functions of an organism that either don't change or that are difficult to change. We could say that these are evolutionarily old and defining characteristics of the organism that need to be maintained, as a challenge to them may result in death. By *flexibility*, we mean an organism's internal ability to deal with environmental challenges. We think of flexibility as having a gradient that represents parts of an organism that can be changed, from easily to not so easily. This generally corresponds to the surface (more flexible, like skin) at the easier end of the gradient and to the core (less flexible, the structural core) at the more difficult end. In general, the greater the flexibility an organism has, the easier a given change is to make in time and energy and the less it will affect or challenge the stable or core functions.

In a way, we don't really differentiate the physical organism as a whole from the central information cycle, in the sense that the entire organism can be looked at as part of the cycle. Therefore, when we speak of *flexibility*, we think that the most flexible parts of an organism are its perception and intelligence, which perceive and plan the handling of environmental challenges. The greater perceptive ability and intelligence an organism has, the less the physical action portion is challenged.

The first priority of the action phase is to hold the current structure together and keep functioning, as well as to update and keep the knowledge library safe—in short, core stability. Some of the most basic processes in life have to do with internal stability and self-preservation, which, of course, also take energy and require change. While we said earlier life's basic action potential is expansionary or interactive in nature, it is also constrained by this need for physicality, predictability, and stability, which allows its expression. Therefore, while organisms are ultimately shaped by their effective environments, they also look for flexibility within those environments that can be used to benefit them.

Organisms read information from the effective environment and take action as well as they can while intelligent internal changes occur, based on a balance between stability and flexibility. For example, a bird that has recognized seasons as a stable effective environmental factor and that flies south for the winter normally wouldn't abandon the ability to use that successful, stable strategy simply because it did not get as cold one year. However, in our view, if similar warm winters are experienced over many years or generations, the bird may very well change its behavior, leading to physical changes, which, over time, can also be passed down. In this example it is perception and intelligence that are engaged first; action and physical change occur only when they are needed. We would consider some migrations to be relatively hard information if it has been repeated and validated over long periods of time. In some species, this knowledge could be more flexible, meaning that their migration memory is more recent and more easily changed or less important.

In general, stable information is not changed quickly, because it has a long experiential history of being effective and has been saved into core forms and functions. Basically, decisions as to how flexible a system should remain are based on use and the energy it takes to maintain it. Expensive, unused flexibility that is perceived as no longer being needed won't be retained, although, in general, the ability to turn functions back on is retained, especially over generational time periods.

Maintaining stable conditions requires less energy than change, so life tends to seek stability. Therefore, life strikes a balance between the stability of holding its own structure together to protect its gathered information and the adaptability of retaining the ability to change where needed.

STABILITY AND FLEXIBILITY

Although stability and knowledge protection are necessities, an ever-changing environment does require some continuous internal changes in order for life to keep functioning. To maintain its internal identity, a life-form must generally continuously update itself because of the changing environment. Life builds itself in such a way as to retain the needed amount of flexibility when and where it is needed, as determined from past experience. Furthermore, flexibility has an energy cost that is weighed against the benefit of using resources elsewhere.

Intelligence is a form of flexibility that operates by making behavioral changes rather than more extensive physical ones and is the first line of response to challenges. Based on new behaviors, including physical activity, action potentials can also build and be held until changes become possible. Changes may therefore happen quite rapidly and may seem dramatic on the basis of one small piece of information (e.g., new nutrients), which, perhaps, enables previous information to suddenly be used in a different way or location. We believe this is also the case during development, when action potentials being passed down encounter more flexibility in levels that may be much more difficult or impossible to change in the adult organism.

DEGREES OF EXTERNAL PRESSURE FOR FLEXIBILITY

In very stable effective environments, life may not need much physical change or generational updating. But when there are many effective environmental factors, change may be possible, and the knowledge bank and perceptive abilities can change more quickly. Organisms can find more knowledge in effective environmental factor–rich environments.

Factors such as time, choice, flexibility, and intelligence may allow some organisms to survive by affecting the selection process. Intelligence is usually the first layer of flexibility, in that it allows the possibility of trying different strategies with a given structure. This allows internal systems to project possible solutions to new demands. Therefore, physical versus behavioral change is an interplay between the central information system and that of the internal units.

In times of maximum need (i.e., when survival is at stake) or selection pressure, the intentional drive is fully focused on finding the best solution to a given problem. Given enough time, effort, and physical flexibility, repeated demands are internalized and changes are made that can also make a difference in survival. The intelligently recognized importance of these situations can result in the reorganization of the current information library to reflect the importance of the changing effective environment going forward, including adding or strengthening action potentials to be passed down.

Flexibility is based on what has been built into the system during evolution and includes the ability to change both information and physical attributes. Generally, intelligence and time are the keys to flexibility; intelligence allows an organism to find a behavioral solution, but it also tends to allow an organism's internal units more time to find solutions, including a wider array of possible solutions to chose from. Physical structures and systems are also somewhat adjustable, and if flexibility wasn't built into a system because it wasn't needed historically, then physical change will be a greater limiting factor. Of course, as an organism strives to survive a given demand, it uses historical knowledge—meaning that it will use what attributes it has available. This is subjective, and an organism doesn't necessarily understand its internal workings and what its internal systems are capable of, from our viewpoint. Therefore, we are not implying that an organism consciously knows that a given system needs to change (the subjective internal workings of physiology are discussed in the next two chapters), rather that systems change on the basis of their perceived demands.

Similar to the perception of other environmental fluctuations, an organism's survival in the face of an important event creates a recontextualization of the knowledge base, whereas not surviving until reproduction obviously means that all accumulated knowledge from that organism's evolution can no longer be passed on. We believe information is saved and knowledge updated based on the importance of the experience, and so, a great need and challenges lead to faster change.

Severe stress may create a wide array of possibilities as to who survives and who does not, and although we say that intelligence is a selectable trait, this does not mean that it is the only factor or even always a factor. Many times, it just doesn't matter—if challenges are too great and, similarly, if the challenges are to slight. Perhaps all individuals will survive or perish regardless of past experience or intelligence. Furthermore, under severe pressure and maximum effort, even widely differing and unreliable strategies may be employed after normal strategies prove ineffective, making survival perhaps lucky. Or, perhaps only a small percentage of organisms survive because only a few have the correct information saved in their knowledge bank to make the right choice. Maybe some are just lucky enough to see another try something that works or in some other way survive by means that have little to do with intelligence. There could be cases in which multiple solutions to a problem lead to a divergence in strategy and direction of evolution.

According to intended evolution, then, when we talk about individuals or organisms being selected, we mean that a series of past and present intentional choices and actions have been selected. Therefore, it is the historical summation of intelligent gathering, organization, and use of information (actions) that natural selection has acted on; intelligent action is being selected, and life keeps getting more intelligent.

CHAPTER 5

CHOICES AND ACTIONS

—

THERE ARE TIMES WHEN INTELLIGENCE CAN MAKE A DIFFER-
ence in survival and reproduction and when it can therefore
matter significantly to future evolution. In the information
cycle framework, organisms express themselves physically as
the experience of choices and actions during interaction with
the environment. The experience of physical action is saved
as new information or as an update of the knowledge library.

Choice and action could be called the *execution phase* of
the information cycle. They could be considered both the
beginning and the end, in the sense that they are where
action potentials can be executed as choices or actions but
also where action is perceived. As was noted earlier, this
cycle is not a linear process with a beginning and an end
but, rather, a simple representation of the perceptual process,
in which the execution of physical activity is an important
factor in the learning process. Therefore, a projection, manifesting as action, results in a different quality of knowledge
than does a passing intention alone, although both processes
are perceived and may be saved to the knowledge library

as deemed appropriate. Actions in response to external effective environmental factors can lead to a robust internal change—a remodeling—of the physiology, which itself is a physical memory. When we use the term *physiological remodeling* going forward, we do so because it is descriptive of what we are talking about; it is a useful image, not intended as a precise depiction. This is not to say that perceptions and intentions that do not manifest as physical action do not matter. Intentions not resulting in action that occur over and over, for example, may eventually lead to action if that is deemed advantageous. Furthermore, any intention or perception could generally be said to affect all physiological processes, depending on its perceived importance.

During development in complex organisms, for example, genetic switches are much more flexible and are susceptible to perceptions and intentions. This is a period during which simple environmental input and both choices and actions affect internal growth as the effective environment quickly grows. In general, information cycle activity that results in physical action tends to be organized differently than activity that does not, because of the system-wide experiences that occur internally. What action is taken and what internal changes are possible depend on the internal arrangement and flexibility of the organism, which are based on its evolution. Simple organisms tend to be able to remodel as well as pass on that information (i.e., evolve) more completely. More complex organisms have to take greater previous evolutionary experiences into account and must recontextualize more information from the past, which takes more time and cooperation prior to making system-wide changes. Furthermore, complex organisms are more integrated, in that their internal life is more interdependent in their interaction with the external environment. Complex organisms also have core functions and structures that cannot survive if they are violated, such as most organs and their functions. However, complex life-forms have access to more energy and, through their shared information cycles (see chapter 6), have greater intelligence, perceptual ability, and flexibility to deal with external pressures and a wider variety of choices and actions available to them.

Natural selection can act on choices and actions; in the words of evolutionary biology, a phenotype is selected. Life-forms that can choose well and carry out the appropriate action are more likely to survive a given challenge, which makes choice (or, generally speaking, intelligence) a trait that can be selected. The choice trait, including the ability to hold more choice possibilities (projections and action potentials), represents more flexibility. Of course, any other factors that favor good choices, including biological processes or beneficial mutations, will be selected along with any choice selection. Choices are made on the basis of projections of what will happen in an organism's effective environment—again, a form of intelligence.

We want to remind the reader that although we used the term *selection* in the previous paragraph, according to intended evolution, this is an interaction between the environment and our information cycle framework. Furthermore, we believe that all relevant internal information or knowledge, including the DNA that will be encoded in an organism's offspring, can change on the basis of the organism's intentions, although, of course, we do not know the details of how this is done or to what extent.

Life chooses to pattern itself internally in the most beneficial way possible to fit the patterns from the external environment. Therefore, what nature selects is the process of intended evolution itself; it selects those individuals that are good at the evolutionary process. The external forces of selection, like the rules of the card game, interface with the internal attributes of the organism through its intelligent choices and actions. Success in the process means that better evolvers are selected. This does not go against the idea that the genome is selected; our view is that the genome represents the most updated and intentional accumulation of information about the organism. Of course, we also believe that intelligence and knowledge, even if they are not represented by the current interpretation of the DNA sequence, are also selected in the evolutionary process.

INTELLIGENCE AS FLEXIBILITY

The intelligence portion of the information cycle process that culminates in life's choices and actions becomes a more important factor to evolutionary outcomes as evolution advances. More accumulated knowledge means a higher level of perception, which leads to better projections of the future and, therefore, to more advantageous evolutionary outcomes. Better projections mean a more stable and efficient use of an organism's effective environment, as well as the ability to increase and manipulate it. We could say that the closer the projections are to actual outcomes, the more able an organism is to control its future, which would lead to more success. As life becomes more complex, it develops as shared perceptive abilities among cooperative systems (e.g., organs in an organism, individuals in society) which allows it to increase its flexibility in the form of centralized intellectual decision making.

However, the universal environment and its (natural) selection pressure remain the ultimate context of evolutionary change. In simple life, behavior choices are few, and they may have relatively little effect on survival rates. At this level, intelligence and the choices and actions that result from it are usually relatively weak factors in evolutionary change. Choices and actions, as well as intellectually derived flexibility, are limited in such organisms, and the traditional version of natural selection is the important factor in evolution. As we have said, evolution involves not only what an organism is but also what it does (i.e., with its intelligence), and simpler life-forms cannot do much to avoid external challenges. Here, selection has very little to do with anything but physicality, because there are few effective environmental factors that actually exert selective pressure while allowing for choices. In other words, a microorganism has a very limited effective environment; it does not recognize very much of the universal environment. In general, its use of intelligence is more limited to internal rearrangement according to its immediate external surroundings, with little of what could be described as intelligent planning to avoid external challenges. That said, we certainly know that microorganisms are very physically flexible internally. Their knowledge is proportionately more about their own internal makeup rather than the external environment. Their subjective knowledge—and therefore, their intentions—has much more to do with their own biochemistry and physiology as these pertain to immediate external signals. This is why, over time, microorganisms are so good at rearranging their insides to fight off direct attacks by antibiotics, for example. How this compares to more specialized single cells and systems in larger organisms is the topic of the next two chapters.

THE SELECTION OF INTELLIGENCE

More complex life-forms have a larger scope of perception in time and space to deal with the external challenges that they face. Additional time to use their intelligence and make adjustments is an important factor in the flexibility afforded more complex organisms. For example, according to intended evolution framework, some portion of arctic populations, such as arctic wolves, facing colder climate challenges had the intelligence and flexibility to make behavioral and internal changes that led to the selection of adaptations expressing those changes. Knowing what percentage of wolves were able to make such a transition versus those that were selected out would help tell us how important internal flexibility was in the transition versus selection.

The more highly functioning the information cycle is, the more precise choices and actions can become, and the more interactive and able to deal with selective pressure the organism can be. At the human level, intended evolution argues that internal factors—primarily, intelligence—can become the predominant factor in evolution, which will be discussed later.

As the perception and projection abilities increase, additional choices and learned behaviors allow more flexibility to deal with environmental challenges. However, an increased perceptive ability also comes with increased maintenance (homeostatic) costs. Specialization increases central perception but also increases local stability and maintenance costs (see chapter 6).

Each new situation is another effective environmental factor that has been recognized and for which the knowledge library must be updated for use in projecting potential outcomes more efficiently. Selective pressure remains, but as the effective environment approaches the universal environment in its accuracy and complexity, intended evolution becomes more important, and selective pressure becomes less about survival and more about efficiency. Advancing perception and intelligence allows life-and-death situations to give way to a pressure to conform an organism's activities via planning to avoid big problems. The same factor in early evolution, which was a life-and-death factor, becomes perhaps a simple challenge with a wider scope of perception. Finally, in humans, intelligence and external environmental manipulation become more important in dealing with external pressure. This can replace the importance of internal physical flexibility. Due to intelligence, fewer choices involve survival in modern human societies.

SITUATIONAL INTENTION

We spoke early on about life's basic intentional force driving interaction with the external environment. With a wider scope of perception, many new avenues of expression arise for the same basic intentions, and an organism's intentional force can be directed toward various situations as they arise. For example, in simple photosynthetic life, the intention to interact in order to obtain energy is relatively limited—perhaps no more than some adjustment to the mechanism for gathering sunlight in a given situation. But in more complex life, this same basic intention can result in many more possible *situational intentions*, by which we mean situational derivatives of an original basic intention to deal with a given need. More complex plants can grow more leaves in sunny areas and reduce them in shaded areas, for example. Animals have evolved to get energy from many varying situational sources, whereas plant variation is limited by the intention to use sunlight.

Situational intentions, like any intelligent process, imply the previous description of the information cycle at work, but in a manner that is focused on a goal or situation at hand. When a new effective environmental factor arises, it is inherently subjective, as we have noted before, and, therefore, an organism's perceived situation is also subjective. More complex life-forms have a wider scope of understanding of the challenge or opportunity that arises with each new effective environmental factor; there are more choices and actions available. Life can take advantage of a new effective environmental factor by formulating projections based on the situation. For a given change, a life-form might have many options that entail numerous different potential actions, depending on the situation and its own internal attributes. For example, when responding to a new predator, an animal may run, fight, climb, cooperate, or try some other strategy. A repeated decision to choose a given projected response (the action potential) and act on it will lead to changes over time, as an organism's internal structures (e.g., organs, which, we argue, have their own intentions) sense the change in their respective effective environments. According to our framework, giraffes would indeed not have long necks if the need was not perceived and the intention to repeatedly reach high food sources was not internalized as knowledge and an action potential. Repeated responses are translated inward through evolutionary pathways. When deemed relevant to a given system, that system will formulate its own action potentials (projections), which, over time, are chosen and can lead to action, such as remodeling themselves. Internal action potentials are also shared with other relevant systems and the organism as a whole.

Any such situation will lead to a comparison of potential solutions on the basis of such factors as what is demanded and the importance, internal flexibility (i.e., remodeling potential), and costs, including energy (see chapter 7). A decision as to the best thing to do, given the situation, will lead to a new and repeated situational intention. For illustration, we might say that some of the evolutionary differences between a wolverine and a rabbit could be due to differing intentions to fight or flee, given the situation of a new predator entering the effective environment.

TIME LINES IN SITUATIONAL INTENTION

More intelligent life-forms with a greater ability to hold action potentials can plan, meaning that they may wait for several factors to coincide over long periods of time before converting a given action potential into real action. Information segments can be held, compared, combined, or in other ways processed, waiting for an action potential to turn to action. More complex organisms can hold more and larger projected time lines for longer periods and have more energy to act, giving their intended actions greater scope and detail. We call this "operating on a longer time line." When conditions are right (new conditions allowing a chemical pathway to change, a mutation to happen, or other choice to be possible), an organism may chose an option that turns a held action potential into an action, leading to a new strategy for a given situation. Therefore, the time line for physical internal alterations is stretched out, as many solutions to the problem can be processed. We do not know the mechanism, but we believe that when time segments represented internally allow projections of multigenerational time frames for action potentials, this information can be passed to the next generation. Long legs and long necks are both possible solutions to reaching high branches, although the change may require a number of generations. Although each generation may experience a somewhat different environment, if a given condition persists, the decision will be made to remodel at some point, probably when it is least expensive, such as during development or when certain resources become available. With birds as an example, differing beak lengths, shapes, and sizes have been studied. We believe these changes also developed intentionally, based on a perception of the need and action to obtain a given type of food. When enough information is available about a given situation (e.g., by repetition), a projection is chosen and changes are made. This activity is not exclusive of natural selection; rather, it accelerates its effect.

New functions and structures are built in the context of the old, and, therefore, more complex life needs to take more historical information into account as its internal structure adjusts to a given demand. Thus, it makes sense that changes applying to an evolutionary time scale need not be built in the generation when the demand first arose, since such changes would not make sense unless the demand is repeated many times and previous knowledge is recontextualized over multiple generations. From this viewpoint, it would make sense that DNA, when described strictly as sequences, could take generations to change, although we qualify this in our belief that much more needs to be learned about the details of what and how information can be stored and passed on. Finally, many physical changes will be made on the basis of resources and available energy. Even though a given structure may not be functionally perfect from our viewpoint, this is what a given organism has come up with from its subjective viewpoint and available resources. The panda's thumb is a very good example of life finding a solution to a very specific situation. Therefore, situational intention can lead evolution in very specific directions or more general directions, it may be saved for the future, or it may simply be abandoned as new information comes in.

THE SCOPE OF PERCEPTION AND SITUATIONAL INTENTIONS

As the scope of perception increases, it results not only in larger physical and temporal horizons but also in more and more detailed effective environments as well. The advent of glass as an effective environmental factor for humans and the subsequent refinement of its use led to the ability to see previously unseen objects (e.g., through a telescope or microscope). Therefore, perceptive scope means distance but also detail. The advent of information about cells and microbiology, for example, has revolutionized the understanding of our environment and evolution.

An increase in scope means that an organism's effective environment takes on new complexity, especially in relation to increased interactions with other organisms (living effective environmental factors). Relationships with other organisms potentially create especially large variations in the immediate recognizable environment. This is because of the complex behavior of living things, which takes time and intelligence to evaluate and adjust to. Situational intentions can also be about what an organism wants to do over time and can be directed more toward a goal. A new and persistent predator in the forest can lead to the long-term intention to coexist via different situational intentions: get faster, move in more evasive ways, sleep in new, safer areas, eat different foods, and so on. Of course, the organism will choose the easiest and most efficient way subjectively, on the basis of its existing talents and attributes at that time. Each of these choices would lead to an organism of different form and function, and evolution can go in different directions. There may also be short-term measures to deal with an immediate threat while better, longer-term solutions are worked on both internally and externally, perhaps held as action potentials for future generations.

Situational intentions, like other intentions, may result in genetic rearrangement and manipulation within the body, accompanied by physical changes if appropriate. All of this depends on the organism and, of course as always, on the properties and importance of the effective environmental factor that presents the challenge.

CHAPTER 6

CORE INTEGRITY
AND COOPERATION

—

AS WAS NOTED EARLIER, STABLE AND RELIABLE INTERNAL
conditions and structural integrity, differentiated from the
external environment, are a prerequisite for life. Furthermore, in our framework, internal conditions must be such
that experiential knowledge about the environment can be
efficiently stored and used by an organism at relevant levels of functioning. As a life-form becomes more complex, a
greater interdependence between its internal structures and
systems develops. Greater interdependence within a group
means greater stability of the relationship, higher maintenance costs (from the organism-level viewpoint) for the
shared environment, but also greater centralized perceptual
abilities. This is because interdependence implies specialization and increased function in the area of specialty. Since
this more detailed information is shared in the group, overall
there is greater perceptive ability of the group.

HOMEOSTASIS AS A STATE OF COOPERATION

Homeostasis, the property of a system in which variables are regulated so that internal conditions remain relatively stable, is one of the generally accepted attributes of life. Like other processes of life, it is, according to the theory of intended evolution, intentional in nature.

An organism's stability is essentially a form of self-preservation, which also implies that the organism must be able to change, because the environment is constantly changing. Organisms must therefore exhibit internal control or self-control with respect to their physiologies. By *self-control*, we mean the ability to maintain the internal stability and predictability of internal environments. This allows for greater flexibility with respect to the organism as a whole in its interactions with the changing external environments. Since perceptive ability is shared, stable internal relationships allow a greater portion of it to be used elsewhere (see Specialization and Cooperation, p. 83).

Although homeostasis is usually spoken of from the perspective of the organism as a whole, even single cells exhibit homeostasis, as do other units of form and function inside an organism. The use of or a relationship with any effective environmental factor implies a benefit to an organism, including effective environmental factors shared with other organisms or simply other organisms themselves. Organisms in beneficial relationships with others are acting in cooperation; those in nonbeneficial relationships are neutral or in competition, either directly or in relation to a given effective environmental factor.

Just as an organism might use inanimate things to benefit itself, such as protection from the elements, so it is with other living things. Two or more organisms may choose not to interact, may choose a win-win interaction (cooperation), a win-lose interaction (competition), or perhaps a mix of these. Multicellular life is primarily predicated on interactions based on the mutual benefit to the individuals and, therefore, the organism as a whole. In this case, two or more organisms (cells) enter into a cooperative sharing of information about effective environmental factors as an attempt to satisfy their basic shared intentional force. In a cooperative state, organisms or the individual cells of an organism share information, meaning that their information cycles are combined on the basis of the nature and degree of cooperation and interdependence.

Although the organism's internal systems and cells are still individuals, they also function collectively and share information. When two or more individuals share information, it creates a cooperative state resulting in the emergence of unique properties not available to them separately. This includes a collective intelligence that allows groups to function as a unit, as well as individually. An internal organ is an example of this, as are other more local groups, or *functional units*, within an organ; this is a similar phenomenon to insect colonies, flocks of birds, herds of animals, and schools of fish, which are examples of cooperation among groups of similar organisms. With reliable, stable sharing of information and perceptive abilities, knowledge building can proceed more quickly at the organism or group level. Evolutionarily, such mutually beneficial cooperation resulted in a situation in which some cells became internalized, so to speak, in return for getting resources delivered and waste products removed.

TIERS OF COOPERATIVE ENVIRONMENTS

Mutually beneficial relationships that resulted in the development of an environment benefitting all cells was a basis for an early form of homeostasis, a progressive cooperative effort, on which further versions were built as evolution resulted in more complexity. Even two cells that attach to each other can be considered as being in homeostasis if it allows them some sort of savings or benefit—not having to maintain certain aspects of the common part of their cell walls, for example. Because there is a perceptive interface, information cycles are shared in some respect, based on the relationship. The formation of groups by cells, organs, or organisms implies a mutual benefit; otherwise, the relationship would break up over time. Current homeostatic systems (and all internal systems), therefore, are a reflection of evolutionary histories' cooperative growth, much like the basic infrastructure and service systems reflect a city's history. At each step, there was an organism (or organisms), with a shared, centralized information cycle made up of smaller individual units. By looking at human physiology, for example, we see many tiers of balanced relationships. Cells can be part of a specialized functional group, meaning that the overall group is part of the environment of each of its members. In turn, the functional group's environment is a larger group—a tissue or organ—whose environment is the system of which it is a part.

From the viewpoint of an entire organism, homeostasis is the maintenance of stable, reliable, internal relationships. This facilitates optimal cooperation and the combination of intelligence, as described by the information cycle framework. In other words, it allows commensurately more accurate external perceptions, projections, choices, and actions from the perspective of the entire organism. An analogy would be the citizens of a country working against a common challenge (such as a foreign army). By communicating to share information (combined perceptual ability) in order to act as a group and raise its own army, all units benefit more from cooperation than from standing alone. This cooperation results in an increased ability for internal stability. Notice that, on the basis of earlier discussion, these two attributes are not separate: Increased external perception (via specialization) is part of internal cooperation and stability, because it allows interaction with the environment at the combined level. The centralized, shared information cycle is therefore the first level of defense for the group against selective pressures. Homeostasis maintains the internal environment for cooperation, meaning that individual needs are met, which also ensures a benefit for the organism as a whole. The internal units in cooperative states help the organism's information cycle to run smoothly, because individual information cycles are not separate from the central cycle when in a cooperative state with it. Each unit has its own "mind," which is also part of a larger mind, so to speak. Therefore, any cooperative breakdown in local benefits—an imbalance—affects functioning, including centralized perception and intelligence. This imbalance may be perceived by the organism as disease, and if conditions are bad enough for the units to decide to try to change their functional relationships, it could lead to a possible breakdown of the system as a whole.

INTENTIONAL SELF-DESIGN

Homeostatic systems were not developed nor do they operate only from the viewpoint of the entire organism in its current form. In other words, we do not imply that an organism is a third-party designer of internal systems. Instead, the internal systems develop locally in a cooperative manner, and as evolution progresses, they reflect the entire evolutionary process of the organism up to that point. The information for the organism-level entity, such as the relationships among its internal entities, is saved all along the way and used for all processes, such as repairs or reproduction.

Using a design metaphor, we might say that evolution is a self-design process that aims to achieve the best fit for an organism to its particular environment. Organisms develop from within, during a process in which a hierarchy of units continuously manipulate themselves to create more beneficial relationships. Because it is the shared perceptive ability of interior life that allows an organism to act as one with respect to the external environment, external interaction is also perceived internally. Any external perceptions leading to a need for internal change work their way down to local levels through the hierarchy of environments. Perception at each level is subjective and always based on the internal knowledge of the unit. Internal units also interact locally within their respective systems and structures, reflecting their internal relationships. This chain of events is based on communication pathways developed during an organism's evolutionary history.

STABILITY AND FLEXIBILITY WITHIN A COOPERATIVE STATE

It is the organism's internal flexibility that gives it the ability to adjust and, therefore, to regain stability—homeostasis—in response to external challenges. There is a tendency for internal systems to seek stability, resulting in some flexibility when dealing with these external challenges. Flexibility is built into any relationship on the basis of the nature of that relationship and can be thought of as change that can be sustained without the relationship collapsing. More generally, we could say that flexibility is the ability to change without losing identity, a natural result of dealing with external fluctuations. This basic dynamic is also linked to the expansionary nature of life's basic action potential, because greater perception and planning ability eases environmental stress.

There is a balance between stability and flexibility. New effective environmental factors can cause challenges to stability, the responses to which require flexibility. An example of internal flexibility allowing for greater external interaction is the potential for varying heart rates, which allow an animal the strategy to run faster or farther under certain circumstances and yet not waste energy when exertion is not needed. Early-warning strategies for predator avoidance are a behavioral flexibility that developed to keep the interior relatively stable at low cost, as opposed to the constant use of flight responses, which require energy and time to return to a balanced state.

Internal systems reflect the flexibility needed to return to stability after imbalances are caused by environmental challenges. Therefore, flexibility is an intelligent process built into a system as it developed and is based on repeated experiences. When experience shows that certain functions are taking up energy but are no longer needed, they may be dismantled, whereas experiences that require certain functions to be maintained will result in flexibility being retained by that unit.

An interesting example of retained internal flexibility by a system over long evolutionary time periods is the function of the bones of some organisms to mend after a break. Greater flexibility also implies greater possibilities for internal adjustments in response to challenges, either to keep the internal systems stable or to change their structure. However, this also implies less specialization and automation and a greater energy expense by those systems. Internal change will be discussed further in the next chapter, which concerns physiological remodeling.

Although internal cells and systems change and evolve in conjunction with the entire organism, their relationship to natural selection through the external environment is subjective, based on the ensuing change of their internal situation or environment when the organism as a whole is challenged. Internal life has effective environments, consisting of its place in the interior of the organism. Evolutionarily, cells that once needed to attend to signals from the external environment as individuals internalized their function through rounds of cooperation and now get their cues from inside the larger organism, including their neighbors. Over time, shared perceptive abilities led to further cooperation to deal with external pressure and signals. When special situations or needs arose, relationships were formed, including specialized functional groups or organs being formed to deal with them. Each new round of cooperation implies that previous cooperation is adjusted but is still in place; otherwise, cooperation would end, and a given form would disband or at least radically change. We could look at these new rounds of cooperation as a layering of new systems over the old systems, such that newer layers still cooperate with or provide a service for the older ones, albeit in an adjusted fashion, since the inner layers change to adjust to the new terms of the cooperation during the process. Like the way life tends to reuse and add to effective tools such as chemical pathways or processes, so are living structures and processes reused when possible or incorporated into a new form when it is mutually beneficial.

SPECIALIZATION AND COOPERATION

Specialization is sometimes defined as the setting apart of a particular organ or unit for the performance of a particular function. According to the intended evolution framework, nature of specialization is that individual cells and systems give up or narrow individual perceptive functions and flexibility to the group in return for increased local benefit, resulting in greater stability. This cooperation creates a more stable environment for internal life to respond to, because the participants each deal with only part of the overall challenges and can reduce energy usage on previously needed functions to focus on a given task. Through each participant concentrating on patterns of information in its specialty, the perception of the group as a whole is greater in needed areas than it would be if each unit functioned alone. Therefore, all individuals end up benefitting through increased combined or group perceptive abilities in return for maintaining local stability as well as group self-preservation.

More repetitive and automatic activities require less intelligence but also require services to be delivered to the local level. This means that overall maintenance costs increase but are outweighed by the advantage of perceptive increase. This is true of any cooperative unit; specialization requires maintenance but yields benefits overall. An organ's cells, for example, are specialized to excel at specific tasks, but they have stopped performing basic functions such as finding their own food and therefore have to be fed by the vascular system. If a cell gives up the ability to perform certain functions, it is because providing more of one thing in return for functions performed by others is beneficial. Therefore, generally speaking, overall maintenance costs increase as evolutionary complexity increases, because increased complexity implies greater cooperation and controlled local environments.

However, from a local perspective, the costs in dealing with (the more stable) local environments fall. What is given up in local maintenance costs is made up for by expanding group perception to deal with environmental challenges as a team. It is not implied that local units understand these concepts as we explain them here; they are simply entering into mutually beneficial arrangements at a given time, which results in benefits to those involved; otherwise, it wouldn't happen. Cooperation allows both internal efficiency and stability and, therefore, increased organism-level perception and projection ability with respect to the external effective environment.

In summation, homeostasis is the result of cooperation that results in a stable, predictable, and flexible internal environment. This cooperative functioning of the group—a centralized information cycle—results in combined perceptive abilities and knowledge collection. This, in turn, benefits all of the internal life of an organism. The current homeostatic arrangement is the most up-to-date version of coexistence and cooperation of internal life for dealing with the current external environmental demand. In other words, the internal systems of an organism are a reflection of its cooperative evolutionary history in dealing with its perception of and interaction with the external environment.

INSTINCT AND DEGREES OF AUTOMATION

Even in humans, there are many behaviors that we would not think of as *intentional*, such as reflexes or instincts. By *instinct*, we mean a relatively fixed pattern of behavior in response to a certain stimulus.

According to the theory of intended evolution, we should look at reflexes or instinctual behaviors as automated versions of what once were more intentional and intelligent decisions made by the organism. Depending on the relevance or repetition of a paired choice and action, a shortcut of the intellectual portion of the information cycle develops, resulting in decisions becoming locally automated.

In other words, a given sequence of behavior at one point in evolution may be represented as follows: A++B++C with the plus signs signifying the use of a given amount of time for intelligent processing; more plus signs indicates that more time and processing are necessary. Through repetition, the interaction transforms over time and becomes automated: A++B++C → A+B+C → ABC.

In general, unrelated functional ability is slowly dropped when it is no longer needed, and repeated information cycle activity of the same input induces specialization and automation to increase efficiency and save energy. This reduces the necessary level of intelligence (required processing power) and the amount of time for building action potentials.

More repetitions lead to more stable projections for the next time, and, therefore, over long periods, less intelligent evaluation would be needed before the action is taken. This is the result of the natural decision-making mechanism of any living system. Energy is saved because of reduced processing time, and the reliability of execution is increased as the time to action is shortened. If the success rate is projected to be better with a shorter response time and the same action is repeated often enough, repeated intellectual decision making may become inefficient, and an autoresponse mechanism may develop: A+B+C → ABC.

Automatic processes such as reflexes can skip thought process at the organism level. This is a clear advantage in fighting or fleeing a predator, when response time is critical. Various levels of automatic responses develop as potential choices and actions are selectively eliminated, which optimizes the function at hand. Automatic decisions such as these are very reliable and add to the stability of the homeostatic system. This is not to say that an automatic response from a system or an organism means that that system or organism is not still intelligent in the moment on the local level. These are living cells running their information cycles like any other life but in a highly specialized manner that has developed over time. They have their own "minds," so to speak, but have trained them on a single task or on a narrow group of tasks with speed and specificity. This means that highly specialized cells or functional groups, such as those in an internal organ, no longer need a large portion of their information cycle (intellectual) capability because of local environmental stability resulting from the cooperative process leading to specialization.

Looking from the organism level, emotions may be examples of evolutionarily semiautomated situational intentions resulting in stable and reliable actions for the entire body to a given stimulus. Fight or flight responses (anger or fear) and reproductive drives are good examples of this. We can consider instinctual behavior as a natural development during evolution to deal with situations in which quick or otherwise crucial action was needed to protect (self-preservation), benefit, or perpetuate (reproduction) the information cycle. We could look at instinct as essentially a series of reflexes, or a system-wide reflex. Of course, we are not saying that this is a black-or-white process; many internal processes, including those noted here, can be seen to have degrees of automation.

CHAPTER 7

REMODELING

—

PHYSIOLOGICAL REMODELING IS A TERM USED TO DESCRIBE THE process by which the body changes, or "remodels" itself in response to a change in a given local effective environment. The reader may have already noticed that we use this term loosely and not precisely as dictated by current standards. We use the term because of its descriptive usefulness but more to mean purposeful internal change in a general sense. *Remodeling* is usually not used to describe physiological actions, which are essentially a continuous process in response to environmental stimuli but, rather, to describe a structural change to the physical makeup of an organism. For example, salivation and coughing are internal changes but not remodeling, whereas a physical alteration such as a lung cell responding to smoking, which usually takes time and repeated exposure to smoke to occur, tends to be called *remodeling*.

According to the intended evolution framework, if a given stress needs a given time period to manifest as remodeling, it is because the cell or system is using its historical context to dictate form and function. Over time, with repeated exposure to (perception of) a new environment, action potentials can build on the basis of these more recent perceived and saved experiences. At some point, the cell or system decides that the original historical context needs to be updated by the new information, and remodeling occurs. Changes such as this can happen locally or at different levels, depending on a given demand. For example, if tissues perceive damage or bones perceive an added stress load, they will act to close the wound or add bone mass. Interestingly, remodeling is pretty much in line with the process that Lamarck described when he said that change is due to need or usage. Of course, the full extent of knowledge about internal activity of this type was unavailable at the time of Lamarck and Darwin.

TIERS OF COOPERATIVE INTERNAL CHANGE

According to the intended evolution framework, remodeling processes in physiological systems should be viewed as being done intentionally and intelligently at some level or combination of levels. All living structures, systems, and processes are intelligent in nature and intentionally attempt to change or adjust when needed. When they do so, it is in response to the relevant environmental conditions, as with all biological change. The details of these processes are complex, but they are actually a basic function and process of all life, done by subjective reading of local information. We believe that observing remodeling as an intelligent process would give valuable input to the cooperative nature of given structures and systems, help describe the scope of communication between them, and provide other valuable information. Furthermore, according to intended evolution, as with other perceptions and actions, any remodeling activity may be important input for multigenerational changes.

We want to reiterate here the brief point made in the information cycle chapter about "layering" or what we are calling tiers of cooperation. Organisms and their internal units can use their historical information to regain past form or function to some extent through communication back through the historical layers. The farther back, the more layers, the longer it would take. Therefore, when we try to observe evolution in nature, including testing the ideas put forward here, this should be kept in mind. We would expect more evolutionarily recent changes (outer layers) to be able to be reversed relatively quickly, compared with core or deep form and function. Therefore, we think of processes that use more intelligence (i.e., that are less automated) as being more flexible and, therefore, more easily changed. These are the same processes that absorb challenges and protect the more rigid core.

We put forward in the last chapter that as life-forms cooperate, share information, and specialize, they change when possible, according to the demands of the local environment, based on their own history. For the organism as a whole, a stable, more automated internal core environment (homeostasis) allows greater central perception and projection through specialization. Of course, greater organism-level perception also benefits internal life by such things as finding resources, avoiding danger, and other problem-solving needs, which is really a form of flexibility for the organism as a whole. These are the type of activities that benefit all the various internal groups and are examples of why life cooperates and shares information cycles.

Flexibility, or the ability to change in response to external challenges, protects the internal core of stability and requires intelligence to deal with the environmental demands that determine the need for and the method of change. In general, the latest layer of evolution is the most flexible and reacts to challenges in a much more flexible way than evolutionarily internal layers. Flexibility, therefore, involves action, which is undertaken to address change in a respective environment. In this way, a behavioral change tends to occur first, using the intellect, and if this flexibility is stressed, greater physiological change is needed. This type of change follows the information cycle framework: An organism's interior flexibility allows changes to fit its interaction with the exterior environment, which, in turn, allows greater interaction, which may lead to further internal changes, and so on. In a subtle way, this is happening all the time in the regular course of events.

For example, in the case of bone tissue, when a local structure or functional unit perceives certain changes in its effective environment (say, repeated high-impact activity or a break), it uses its historical internal knowledge base to make changes that are projected to be successful in returning to the original plan. Cells may break away from their previous relationships and change form or function in accordance with a new environmental need of a larger group of which it is part. Some functional groups, can access internal knowledge as to how they originally developed, which can be referenced during an event such as the bone break. Evolutionarily, healing functions are presumably structured—not unlike instinctual behavior—to be activated on the basis of the recognition of certain events. Like instinctual behaviors, these automated responses developed intelligently and operate on the basis of the relevance or repetition of challenges. Interestingly, bones can remodel slowly over time because of repetitive stresses or can act on the basis of only one event if that event is relevant enough, such as a break.

DEGREES OF REMODELING BASED ON PAST KNOWLEDGE

These previously stored strategies are also an example of the retention of flexibility that may be needed in the trade-off between stability and flexibility. Although they are highly specialized, bone cells evolved to retain the ability to change form and to move, under certain conditions.

This function is still available because these events were historically important enough or were perceived often enough that it was not beneficial for local cells to give up the ability to recreate the original structure on the basis of historical precedent. Note that although a person would certainly experience the bone break, situations such as these do not necessarily entail the central information cycle, signaling details down through the evolutionary hierarchy. The reactions may be regional or local, depending on the situation. The local groups may deal directly with challenges; such is the case of a minor skin wound. Local or regional groups normally have some (subjective) awareness of the central cycle, depending on who they are and the situation. The central cycle, however, has the function to represent the whole and is often unaware of local conditions, presumably based on evolutionary relevancy of communication needs to the central system (see chapter 16).

DEGREES OF INTELLIGENCE IN REMODELING

When we say that all living systems are intelligent and exhibit intention, it does not mean, for example, that each step of a chemical process is intelligent. Rather, we mean that a given life-form uses these processes intentionally, just as humans put water in a tower on a hill. Life makes decisions to use nature, including chemistry, as a set of tools. Physiological processes and anatomical structures are affected by how any organism lives its life—in other words, what it intends to do and what actions it takes. In this way, life mirrors its environment, because its form and function are based on what it has encountered or experienced.

Many aspects of the differing life-forms that we see today are related to life's choices and actions during its interaction with its effective environment. Although the universal environment is the same, differing experiences and knowledge bases mean subjective experiences (different effective environments), which, in turn, lead organisms or internal functional units to go in different directions. Therefore, environmental factors induce biological change subjectively since different signals are read based on historical knowledge. This fits the theory of intended evolution, because biological change is intentional, the natural result of perception and interaction with life's surroundings. In summary, we may say that although life is open to all of the universal environment's possibilities, an organism's evolutionary history also dictates what will be projected to be the best way to proceed.

CHAPTER 8

ANTICIPATION AND PROJECTED INTENTIONS

—

THIS CHAPTER AND THE REST OF PART ONE TENDS TO PRESent in terms of what we have been calling the "central information cycle" and its interface with the external environment. We want to remind the reader that, although we tend to speak of the brain in relation to the central information cycle, of course, like other physiological layers, the brain is also correspondingly layered and therefore important for internal aspects (e.g., automated functions) as well.

As we have said, information, or knowledge, implies a sequence—a recognizable pattern over time. Therefore, knowledge implies a temporal component, and its usage implies some ability to make projections about the future and memory of the past. As humans, we tend to think of information in a very elaborate manner, but the basic ideas are simple. For example, if an incoming pattern is perceived as ABCA and a pattern of ABCABCABCABCABCA has previously been experienced and saved, a projection that B will follow next can be made.

Patterns larger in scope incorporate and manipulate the simpler ones. Darkness turning into daylight and back again is a longer sequence of events or moments that can then be compressed into representative chunks such as a day, a week, or a month and manipulated and used in that way. When something is experienced over and over again, it is recognized as a pattern that is useful in predicting other events in a given environment, such as the arrival and departure of other organisms in a certain location. Therefore, the intention to do something is based on information and implies a predictive element. For example, an organism sensing danger is reacting not only to that moment but also to what it projects to be coming on the basis of its knowledge of past experiences. Not only are the perceptions of immediate demands important to survival, but so is the knowledge on which intelligent projections are based. The more reliable the knowledge used is, the more accurate and useful the resulting projections about the future will be. Similarly, the more intelligent the organism is, the better the use of its entire information cycle framework will be in the choice of a projection to act on. As an example, good maps (better knowledge libraries) lead to better projections about where one will end up in the future than do faulty ones—as will superior intelligence to understand the usefulness of the map you have.

Some organisms may make projections about events farther in the future or more geographically distant than others do, some will do so more accurately, and some project more numerous or varied possibilities. We call these distances from an organism's current place in time and space its projective or perceptive *scope*.

As was discussed in chapter 3, differing organisms have perceptive abilities of varying scope. As life becomes more complex, it accumulates enough knowledge and contextual flexibility to permit holding more patterns in mind that can create more elaborate projections of future environments and events. This is an outgrowth of increasing perceptive scope and allows the ability to see more clearly what is happening and to more accurately make projections about future events. Compared with an organism that may have to physically act out every scenario to experience it and learn about it, the ability to project and even experience through mental simulation (see chapter 9), avoiding any physical risk, would obviously be advantageous in evolution and a selectable trait. This gives an organism the ability to choose from many and more accurate projections without having to risk doing them physically.

INTERNAL SIMULATIONS AND THE DEVELOPMENT OF CONSCIOUSNESS

Conceiving simulated versions of the effective environment may be a response to the demand for much more sophisticated interaction with living effective environmental factors that became more prevalent as life became more complex over time. The demand for projection and prediction is much greater when dealing with living things, which increases the demand for the expansion of information-related specialization. As greater and more complex interactions occur, more processing is needed since more related patterns are filed, grouped, manipulated, and saved as knowledge. When dealing with a living environmental factor, to demand even the same degree of predictability in a given environment means an increase in processing power, or greater intelligence. Various scenarios can be rejected as certain possibilities are ruled out or accepted conditionally so that they can be compared with other conditionally accepted scenarios.

As a general framework for intelligence, any number of information patterns or time lines may be held in mind in relation to one another on the basis of many factors, including etiology (e.g., sound, vibration, vision), relevance (to the current perceptive context), current intention, and so on. The entire information cycle may run over and over without the actual physical execution portion, including the choice of the best projections or the best parts of some to use with the best parts of others, and so on. The process then repeats, increasing in accuracy with each iteration. Ultimately, if action occurs, the results are also stored as updated knowledge. Running scenarios is a significant advancement in evolution and is driven by ever more complex effective environments as perception expands.

With centralized scenarios running, the information cycle can operate mostly on the basis of knowledge-based projections rather than a moment-by-moment link to a current perception of the external environment. Action potentials based on projected scenarios can lead to advanced forms of projection that we could call *anticipation* or *planning*. Humans, for example, can plan an event well in advance and revisit the idea many times before execution occurs, slowly collecting the necessary information from the environment over time when opportunities arise. Furthermore, this could be a shared intention or group effort (e.g., a business project) as well. We take these things for granted, but long-term planning (intentions) can change or set an intention that affects all other ensuing incoming information until the action potential is either executed as action or abandoned.

Intelligent and intellectual activity such as this mitigates environmental challenges, because preparation can be made to meet those challenges. This type of intelligence stretches the response time line to a given pressure and is an example of an increased outer-layer or intellectual flexibility while lessening the need for more expensive internal layer flexibility in response to challenges in general. Animals that prepare for winter, for example, save energy by avoiding having to find food during the most energetically expensive times. Scenarios imply action potentials based increasingly on projective content, although the original context is still an intention based on the organism's effective environment. Advanced projections such as running scenarios have to involve some reference to the organism within the scenario for them to be useful. As this type of mental activity evolved, what we think of as *self-awareness* also began to emerge.

CHAPTER 9

HUMAN INTELLIGENCE

—

WHEN DECIDING AMONG MULTIPLE CHOICES OF ACTION with the limitation that only one choice can be made, making the best choice is the obvious intended goal. Repeated attempts at actions can be limited by time and resources. When projecting an escape plan strategy, for example, one cannot run, climb, and swim at the same time, nor can one afford to attempt pursuing all options available. But, when possible, one can project the outcomes of all of these actions and choose the best one. On the other hand, if an action is localized or detailed enough, like the use of fingers and hands, repeated physical attempts are not so constrained.

VIRTUAL ENVIRONMENTS

Tangible physical action is limited by time, resources, and physicality, but informational manipulation using intelligence is not nearly so. Speaking generally, mental planning eases selective pressure on the physical body, although natural selection is ultimately still at work shaping this activity.

Running scenarios intellectually accelerates the evolutionary trend of handling challenges with the relatively flexible outer layer—the intellectual part—of the central information cycle and avoids the relatively expensive internal, more physical changes. Humans are able to project many choices and to run quick mental scenarios to see which one is best. We can create and perceive scenarios based on experience to deal with our effective environment. Scenarios are based on previous experience, and this activity—processing—occurs at varying levels of awareness as far as the central cycle or perception is concerned (we are speaking generally here and not making claims pertaining to psychology or the detailed workings of the brain). Furthermore, scenarios may have little to do with the immediate physical effective environment. We could call this type of mental activity a *virtual environment*. Note that this label is for discussion purposes only and is not being proposed as separate from other mental processes but, rather, as a description of a relatively advanced use of intelligence. By definition, this advanced use arises out of environmental demand. In other words, it is the increase in information input that drives intellectual demand. The increased manipulation of information patterns (knowledge) about the external environment, interfaced with those from and about the nature of the interior of the body, led to the need for increased processing power, resulting in what we are calling virtual environmental use.

This is where intended evolution begins to differ markedly from the idea of natural selection. It is especially in noting the human ability to create a virtual environment that we can see the need to formulate a broader theory of evolution.

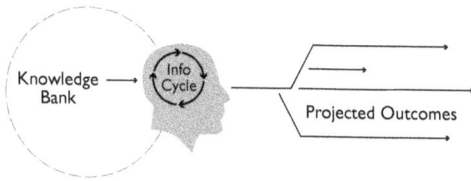

FIGURE 7—VIRTUAL ENVIRONMENTS
Experience, stored in the knowledge bank (memory) is processed through the information cycle and results in many potential actions and outcomes.

Humans are increasingly interacting with environments that they, themselves, have helped create and not just in their physical surroundings but also held as action potentials or projections (in the mind).

An exhaustive description of a virtual environment is not possible or warranted here, but as, an example, consider a stable backdrop representing an actual external effective environment over which the behaviors of certain organisms and situations are projected. This includes what we think of as mental work about problems, planning, imagination, and so on. Virtual environments are a way in which knowledge can be used to project scenarios as almost the exclusive input to current perception. In other words, humans can form virtual environments that, although they are grounded in actual environmental surroundings, may have little to do with them. An example of this is simply being in your thoughts and figuring something out about an issue back at the office. This type of activity can dramatically expand potential choices, eliminating poor ones in preference of good ones and increasing the chances for successful actions through detailed analysis. Until now, it has been assumed that perception was closely associated with the surrounding effective environments, but a virtual environment does not necessarily have to be related to the external effective environment of the present moment. Planning for future events may have little to do with one's external effective environment, for example. The use of the virtual environment expands projective accuracy greatly with more time for intelligent analysis through expanded mental manipulation of internal information or knowledge.

A RAPID INCREASE IN INTELLIGENCE VIA THE HANDS

In recent chapters, we have stressed the mental aspects of informational input into the evolutionary process. However, relatively recently in evolutionary terms, physical detailed tactile input and a rapid feedback loop, combined with mental focus (e.g., creating or working with early tools), also developed. Rapid, flexible, relatively detailed physical manipulation with the hands created quick rounds of development. As detailed manipulation with the hands developed, solutions increased rapidly, and processing power was needed. Detailed physical use of the hands is not unlike what we have been describing as it pertains to the speed of mental processes. We believe that it is the rapid, varied, and repetitious physical informational input over long time periods via the hands that helped to build the modern human brain.

PLANNING AS AN EXPRESSION OF ADVANCED PROJECTION CAPABILITIES

Even a seemingly simple example of advanced planning reveals great complexity when examined closely. If you plan to do a certain activity tomorrow, you may hear the weather forecast and quickly run scenarios on the basis of what clothes you have, what your plans are, who you will see, how you are going to get there, and so on. There may be multiple choices for multiple, interrelated effective environmental factors, which can result in many possible solution scenarios. Given the forecast, should you go by car, bike, or walk? If you are planning on walking, what shoes or boots do you choose, given the forecast? Which route is fastest? Who are you meeting on the way? Who is bringing what? What needs to be packed? Furthermore, this held action potential, or intention, is constantly updated as new information comes in, and we can see that much processing power and a very large knowledge library are needed for these types of activities.

Planning can be done beforehand for a trip, which creates a much more stable and reliable situation going forward. When plans have been made, perception can revert from the virtual environment used to create them to the current effective environment as the action is executed. Furthermore, increased planning time (virtual environment time) can afford an increased projective distance into the future. People plan detailed vacation activities months or years in advance and even start planning for retirement and death decades preceding the actual events.

Planning is taken for granted but is very important. This type of activity is the result of increased time line capacity being saved in the knowledge library and allows increased time between projection and action. By this, we mean that patterns have been put together to project outcomes into the distant future. This means that the intelligent, mental, or intellectual portion of the information cycle is used to ease the demand on the action portion, which is more related to energetically expensive activities, including structural change. Internal structural change also takes a lot more time and energy to accomplish than do behavioral changes. If we look at evolution as a gathering of information, this means that evolution is much more rapid when it includes virtual environments.

Finally, knowledge can also be gained from a virtual environment. Simulated experiences allow projection about complex factors such as the behavior of other organisms. For example, accurately predicting other people's or animals' actions for given situations is of immense value in creating future stability. This is not to say that you are skipping the physical step. Rather, what you are gaining is based on previous experience, and in virtual environments, you then experience recalled knowledge. Virtual environments are useful not only for including more numerous and varied effective environmental factors in making plans but also for increasing the resolution of the details of existing effective environmental factors. Just as understanding the composition of the components of a building allows projections about its future, knowing how the interior of the human body—or even a single cell—works is of tremendous value in understanding the entire body. It is projections based on such detailed knowledge that allows long-term experiments to be undertaken, for example, increasing knowledge further. These things are taken for granted because we function this way naturally, but if one stops to think about our mental processes, it becomes clear that we "virtually" rearrange information about past experiences and information all the time, which results in learning without physical experience. We want to remind the reader that, although we continue to present our oversimplified framework going forward, we are not ignoring the vastly complex workings of the brain. For example, although we have ideas on the topic, we don't have the time to discuss brain functions such as the subconscious processing underlying conscious thought.

SELF-AWARENESS AS THE DEVELOPMENT OF A VIRTUAL AVATAR

In order to do these test runs on an informational landscape (i.e., to operate in a virtual environment), a sense of self (so to speak) developed as a representation of the participant in the various scenarios. This also creates a differentiation of the self from the rest of the virtual environment, including other living entities, not unlike the avatar in a video game.

The information patterns used to create a virtual environment result in a representation of the self, which leads to the perception of oneself, or self-awareness. We presume that this is the natural result when enough patterns can be held in mind to create a virtual environment, the running of multiple future projections of scenarios. The scenarios are also driven by the intention of the self (the central cycle), and so self-awareness results. This results in thinking such as, "If I were to go out for coffee . . ." As desire or intention arises, the information related to it is gathered and evaluated, and a hierarchy of decisions are made. The brain constructs a virtual environment consisting of all of the predictable situational effective environmental factors and strings together the information in various ways to select the best route to the coffee shop—or whether to go at all.

The concept of a self also allows for the superposition of oneself into another's situation, allowing greater predictive ability. Putting oneself in another's shoes—or, loosely, *empathy*—allows predictions to be made about another's activity. In our opinion, this capacity is a natural result of the ability to save and manipulate large amounts of information to project likely outcomes.

Although empathetic activity is normal and can, like all subjective activity, theoretically relate to anything, it would be more functionally relevant to individuals with similar effective environments, such as the same or a similar species or, at least, very important individuals. This would imply that they share many effective environmental factors or a connection to each other. For example, a human shares few effective environmental factors with a cricket, which makes superimposition less useful to an adult human with a lifetime of experiences and a specific sense of self. That said, empathy is a natural, evolutionarily useful automatic strategy and may therefore be subconsciously used in many situations.

Children, for example, often seem able to more easily empathize with other organisms and even with inanimate objects.

Improved approaches to dealing with external pressure and the maintenance of inner stability were the underlying drivers for the development of consciousness and awareness, in our opinion. The ability to operate in a virtual environment allows humans to quickly run potential options for action on a virtual landscape in the mind whenever a pressure arises.

The ability to run scenarios in the mind and the development of self-awareness allow for readily accessible virtual environments in which to test alternatives before making a choice involving costly or dangerous physical action. Multiple projections of simulations and combinations of simulations can be made, allowing extremely detailed planning over long time periods. Humans can make multigenerational plans—for example, for space travel. Virtually infinite possible scenarios, all based on the internal information accessible to human consciousness, have resulted in the human creativity that we see in the world today.

THE ROLE OF INTENTION IN ADVANCED INTELLIGENCE

Of course, all of our previous discussions of evolution underlie these human functions. As with other evolutionary processes, the central information cycle in humans is built on top of (so to speak) and based on previous versions. However, intention still drives the perception of the environment and the rest of the information cycle, including storing relevant information and additional increases in perception. By this, we mean simply that it is generally what one intends to do at the moment that results in experiences being saved internally, even though one does not choose every experience. Of course, we don't have complete control over our environments, but increased intelligence over time has resulted in fewer and fewer situations that challenge the core functions (life and death).

As we have said, internal units have their own information cycles linked to the central information cycle. This newer scenario-running function is layered over the previous informational processes historically tuned to the memory based on evolutionary history. Furthermore, there is still communication both ways, from the exterior to the interior and from the interior to the exterior.

The human body is still constantly remodeling itself on the basis of intentions, perceptions, and actions. However, these remodeling factors, which previously interfaced more directly with the environment, now have a new input: the results from running virtual scenarios.

Therefore, the internal systems will react to a projected scenario, not just to what is really in the immediate external environment. A simple example is thinking about your favorite food; the relevant internal systems react (the salivary glands are activated), regardless of whether the food is actually present. This brings about an important characteristic when it comes to human evolution: We have an important role in creating the environment that the body evolves to live in, both physically and virtually (see chapter 16).

CHAPTER 10

CHANGING THE RULES: HUMAN EVOLUTION

—

HUMAN SURVIVAL RATES AND REPRODUCTION IN MODERN human society today are not as related to traditional natural selection as they were in the past. The average human life span is increasing globally as altered living conditions have made death by traditional (not human-made) environmental challenges rare in many societies. Presently, humans have no natural predators, have created external systems of climate control, and have mitigated many traditional disease threats through improved sanitary conditions and medicine. As a result, in many societies, traditional environmental pressures have little to do with living long enough to reproduce. Deaths due to accidents and wars are unrelated to genetic makeup acquired through evolutionary experience, and modern life-style choices are only distantly related to traditional environmental pressures. Essentially, living to reproductive age or not has much less to do with the genetic makeup in humans today than it does with other factors.

According to the theory of intended evolution, this does not mean that evolution stops. As long as there is new information being recognized, evolution will continue. However, the effective environments that many modern humans interface with are increasingly human-made; we have the ability to chose and even to help create what environments we interact with. Human beings are still processing information, making choices, and taking actions—and, therefore, growing their information cycle and evolving but also collectively choosing its direction by creating the environment we interact with.

THE INFORMATION AGE

Presently, physical challenges to human life span in modern societies are being buffered by external factors (tools and infrastructure) that create environmental flexibility for humans. Much of physical evolution is being redirected to interfacing with external tool inventions in order to respond to environmental demands. Moving into modern, indoor living conditions, for example, is a form of (external) homeostatic flexibility that replaces living outside. Humans in many societies are facing constant situational demand for mental processing power along with a dramatic fall-off in other physical demands. Intentions—and, therefore, evolutionary direction—are increasingly affected by human-made information, which increases the demand for mental processing, including internal virtual environments.

This is evidenced by the use of modern external tools and time-saving technologies allowing more and more time to be spent on activities such as mental planning (projection).

This external "technological tool layer" increases the reliability and speed of the actualization of intentions, as well as increasing the available information forming the basis for those intentions.

For example, compare the intention to go from New York to California today with that of 200 years ago. First of all, few people would have even entertained the intention due to a lack of information about such endeavors. Today, modern specialized technological systems take care of most aspects of the trip, including planning details, needed preparation, and the actual movement from one spot to another. This type of modern automation results in a vast increase in the probability that the projected intention will be realized—and in a fraction of the time it would have taken before. This frees up time to plan more experiences even farther into the future—for example, the meetings that we have once we arrive.

Intended projections actualize in a much more reliable, predictable, and flexible way (we can just take a different flight if our intended plane does not arrive). Compare this with overland trips taken not so long ago in which the physicality of the traveler was much more involved: A person had to spend months walking or had to acquire horses and wagons for the trip, along with enough supplies to get to the next reprovisioning point. These trips had an entirely different context because of the time they took and the fact that communicating with someone at the destination was usually not even possible. For this same effective environmental factor (distance), a fraction of the time and physical activity are needed today.

The external tool systems used by developed populations

allow human activity to be based primarily on abstract concepts, because the physical needs of the human body, for the most part, are met. The traditional or historical selection pressures of even the recent past are much less relevant and, therefore, today's perception, intellectual intention, and action are not directed toward them. Education and knowledge are the main intentional focus for the average person in the developed world, as opposed to digging deeper wells or planting and harvesting crops more efficiently, for example. Although still related to physical needs, modern actions are primarily based on projected advantages in society, which may be far removed from traditional factors or demands. Furthermore, much virtual environment work could be primarily abstract, as in the study of science and mathematics.

THE CHALLENGES OF MODERN LIVING

The introduction stated that the effect of the internal (intentional drive) side of evolution is especially prominent in more advanced life and that the external selection side less so. The human brain is a reflection of this, and this process will continue as the modern demands on the human brain reflect an effective environment high in informational variables. Other systems and organs will increasingly provide for the needs of the brain in modern societies with other functional demands down sharply. Maintenance of humans' advanced mental abilities requires energy, as well as a flexible network of interdependent systems, such as what we see in human physiology. It would appear that many of today's modern humans need only minimal physiological upkeep for the purpose of running the intellectual portion of the information cycle, which has become the predominant environmental demand. As the information cycle demand shifts to information processing and away from physical action, information about these changes is being recorded and passed down as action potentials, and when internally deemed to be the new reality going forward, we can expect further changes in the human body.

Changes are natural, given environmental variability, and we are not saying this is a problem in itself, but the rapid transition to this state of affairs may confuse or stress the flexibility capacity of the more internal layers of the physiology. Flexibility implies time to make changes, and if demand for change is too quick for the current level of flexibility, any system becomes stressed and, under severe conditions, breaks down. Human physiology reflects an evolutionary history based on an entirely different effective environment than many people in modern societies perceive today. Modern humans have seen their effective environments change radically in the last 100 years. Furthermore, with the recent advent of electronic media, information input has increased exponentially. All new recognized information creates internal change and updates, using flexibility capacity in order to do so. Much of the information processing system we have inherited through our evolutionary history was intentionally built for lower volumes and different types of information that what is being encountered today. This includes what we think of as sensory information processed by our brain but also comprises evolutionarily unfamiliar compounds in the water, air, and food. Any compound that is not recognized or that is considered unreliable can stress internal systems that are evolutionarily relatively fixed and normally don't change quickly, such as semiautomatic or automatic systems. Even if these compounds are familiar to us in this lifetime on the organism level, such as food additives, they may force attempted internal change if they are not recognized at the processing levels. In other words, they may not necessarily be processed optimally at the local level, even though, on the subjective level of the human as a whole, they are well recognized since we grew up with them. Many digestive functions, for example, are relatively well tested and well established and may lack the flexibility to change quickly, since they have seen consistent signals throughout evolutionary history. The human body is a reflection of evolution, and unrecognized information takes time to become recognized—sometimes many generations, depending on the layer of the system that needs to change, as we discussed earlier. To recognize unrecognized information, internal changes need to happen, and systems must be updated, also putting stress on the system based on what layer or age it is saved in. It is also important to remember that more difficult to change layers or systems are less flexible for a reason: Long periods of experience are imbedded in the deep layers of the human body and are deemed very reliable and are, therefore, difficult to change on purpose. Therefore, it is potentially the case that any compound without evolutionary history as it pertains to humans can challenge different layers of flexibility and can be expensive to adapt to, if that is even possible.

HUMAN-MADE ENVIRONMENTS

The idea of evolutionarily unrecognized compounds can also pertain to perceptions of informational content, although as we have noted, the knowledge or perceptive ability of humans is generally a more flexible way of dealing with environmental changes than physical changes of deeper, internal, homeostatic systems. Using modern technology, in general, reduces the demand for the physical body to make changes when dealing with external demands. Technology reduces challenges to the stability of core physiological functions; the relevant flexibility factors are essentially outsourced to technology, which then deals with variability and techniques to gather information through modern technology can augment traditional knowledge and greatly expand the external flexibility available to humans. Increasing the ability to intentionally seek out useful information and knowledge opens up the potential for rapid information cycle growth and choices for evolutionary direction.

In the past, however, human perceptions of effective environmental factors were relatively simple by today's standards, as they were based on a relatively stable effective environment—what we might call *nature*—compared with today's quickly changing surroundings. Evolutionarily speaking, humans have evolved to pay close attention to relevant information from the environment, and the environment has changed dramatically to include a technological layer of information. The potential inputs of human-made technological environments dramatically increase currently available incoming information and the related choices and actions in modern times. Furthermore, these inputs change rapidly, potentially creating unfamiliarity for internal systems that are geared to past evolutionary time segments that were typically longer than today's information supply allows.

These types of perceptions can signal the body's remodeling systems to try and keep up, creating uncertainty and stress if attended to and internalized. This is a potential downside of the increased volume of typically perceived information in intentionally influenced environments today and points to information cycling more important than ever in optimizing

Our instincts and emotions, are based on information from an evolutionary perspective that, compared with today's potential inputs, is often only a distant derivative. Modern problems can be confusing to the physiology, and many questions arise as to how information-based evolution affects our bodies.

CHAPTER 11

THE LIFE CYCLE: WHY DO WE DIE?

—

QUESTIONS ABOUT DEATH ARE OFTEN RELEGATED TO PHI-losophy or even ignored, because death is thought to be inevitable. In a way, treating it as inevitable makes intuitive sense, because, historically, it has been observed repeatedly to occur at certain ages and is commonly accepted. Furthermore, as we have noted previously, in an ever-changing environment, any entity (a relatively stable form in a sea of change) will change over time.

At some point, what appeared as one thing is now different; we could also say that the original entity no longer exists. Over time, a rock becomes sand. What we observe at any moment is simply a phase of that particular formation in a chain of continuous change.

Furthermore, there is a point at which a defined entity first fits a particular definition—a time in which it becomes what it is—as well as another point at which it ceases to fit that definition as it changes into something else. In this manner, all things, alive or not, have a beginning and an end. There is the concept of *entropy*, the tendency of everything in the universe to approach randomness, for example.

Of course, this does not answer the question of why we die, although the underlying tendency of needed change to avoid the loss of structure is, in fact, what life deals with, as we mentioned in the first chapter. However, as has also been noted throughout history, life is special in some ways; life has the ability to hold its form together, to stay integrated, at least for a period of time. Furthermore, life can also add to itself, can integrate new information, seemingly in defiance of this underlying tendency to decay. Therefore, perhaps the more important question is *Why does life stop integrating, or holding together?*

In earlier chapters, we said that life adheres to the universal laws of nature, and eventual decay and death is an example of why it would do this. The ability to integrate or organize is augmented by doing so in the cheapest, most efficient way, and it is much easier and less costly to sustain life by following and using nature's laws than by resisting them. Life is informational in nature and uses information as a means to affect its own future, but if a life-form spends its available time, energy, and flexibility fighting the universal forces, it will be much less effective at obtaining new information. Therefore, the efficiency of following universal forces augments core stability and greater external perception.

INFORMATION USAGE IN SEGMENTS

For our purposes, information is the representation of a pattern of behavior over time. The value of information, or the way life uses information, is to extrapolate patterned behaviors beyond the time frame represented by a saved segment of information.

Say that we discover that when lightning is observed at close proximity, thunder will shortly follow; we can then predict or project thunder if we see lightning. But to be useful, this pattern must be experienced, processed, and then saved as a representational segment—or, we could say, it is *condensed*. The entire event was not saved in this example; information about the rain or other parts of the storm were part of the experience, but the consistent, reliable segment—thunder following lightning—was noticed repeatedly and saved. Exposure to the right length of patterns can provide significant information about the future and the past. As we have said, life builds itself on the basis of useful patterns.

Because we know that life needs to keep its structure to survive, the first information it needs to save is about itself. It needs to know about its own structure and formation in order to maintain it for extended periods of time in an ever-changing environment. Having a condensed representation about a structure will make it possible to maintain or repair it, as well as the framework to allow the additions of new ones. Condensed information is obviously easier to maintain than the larger and more complex structures of the body it represents.

Condensed information is also easier to use in satisfying the basic expansionist intention driving life's force on a generational time scale. DNA, for instance, is the condensed knowledge of an organism and its entire evolutionary past, from which it can reform or repair parts of itself or its entirety through reproduction. Duplication may be an early basic intelligent step that life took to use and control condensed information to satisfy its basic expansionary intention, along with saving information about important experiences. Therefore, two basic strategies that have been employed in fulfilling life's basic intention are the following: An organism can create more of itself to expand and collect a larger amount of environmental information with more units, or it can maintain its current stability and collect it for future generations, including by replication, and pass down the information in that way. Of course, the two strategies are not mutually exclusive, and some combination of the two is what we think happens in actuality.

THE COST OF STABILITY IN CHANGING ENVIRONMENTS

Of course, life-forms must still naturally change over time, just as environments change in nature; interaction equals change. The main challenges faced by an organism are sustaining a stable core structure and an ability to interact and change while expanding outward to occupy more space and time (i.e., in fulfilling its basic drive) in the outside environment.

How life balances these two seemingly opposing goals or tendencies of expansion versus self-preservation depends on its internal structure and the external environment, both of which dictate the interface and potential interactions. If there are few interactions, there is little pressure to change, and an organism could theoretically sustain itself for long periods without breaking down. For example, if the (effective) environment is extremely stable, some life-forms do not evolve very much at all, such as some trees, lobsters, and turtles. If there were no effective changes and the life-form could arrange its structure to live without changing itself too much or too often, it could potentially maximize its current life span.

It is when the external environment is not stable—when an organism has to struggle to survive—that the most change can occur. Change is the key to life and death; adjusting to or managing change results in choices leading to a given life span. Let us discuss some factors at play that lead to the necessity and process of a life cycle or life span. Change requires energy that must be gathered and flexibility, which is limited in complex organisms and cannot be easily enhanced, because their structures are based on their (relatively long) evolutionary history. Normally, rapid change in one generation may be prohibitively expensive energetically with regard to flexibility. This is why the context of complex life is change occurring over many generations (i.e., an animal does not bother to make big changes just because of one odd winter season). This may be one reason that, for a long time, people thought that organisms did not change and that evolution did not happen.

LIFE SPAN AS SYSTEM UPGRADES

One analogy of physiological flexibility is hardware and software in computer technology. As we have said, the physicality of the body, including DNA, is similar to tools that can be crafted depending on need. A computer is made to run various programs for the user on the basis of his or her needs, but if a need arises that requires something outside of the current capability of the machine, an upgrade to the next generation may be required. Software adjustments and some hardware adjustments may be adequate, but, eventually, you may need a new model of the hardware. Similarly, life first deals with environmental changes via intelligence and behavioral control (the software). When it is pushed to the limit of that, hard structural changes may be required (the hardware). Furthermore, a point can be reached in which this may not be possible except in subsequent generations.

For example, if a population eats apples, intellectual solutions (the most flexible response to a demand) are used first: Eat an apple from a low branch, where they are easily reached. If those are not available, more physicality comes into play: One must jump or find another way to reach an apple on a higher branch. If the person has never jumped before, learning to jump is also a software solution with some minor hardware adjustments, assuming that the structures (legs) are there.

If, however, the person must reach higher than current

capabilities allow and one cannot climb and so needs to grow in height, it would require a hardware solution. If all of the apples were on higher branches and if apples were a major source of food, over time the projections could indicate that all apples in the future would be in higher places. In this case, a change of hardware to match that environmental situation may be deemed vital, going forward. However, as you probably know, there is a limit to any type of body remodeling; although our exemplar person can increase his muscle density or improve his jumping technique, the ability to lengthen his legs in one generation is limited.

Other factors come into play here as well, of course: A life-form may have other demands and may not put all of its effort into a single intention. Seemingly random factors such as disease, age, and environmental stress may overwhelm its attempts to survive. On the other hand, if conditions allow, organisms and their offspring will find a way to make the changes going forward in order to survive. Of course, some demands are too much for a life-form to handle, and its inability to change may lead to an inability to survive. We noted earlier that informational changes (intellectual or soft solutions) are easier to make than physical changes. Here, we are discussing a case in which not only was intellectual change not enough, but the available physical changes were also not enough, leading to mortality. According to intended evolution, because information—including multigenerational information—is collected and passed down, there is another dimension to intellectual flexibility: Death is an option, not always a detriment.

STRATEGIC MANAGEMENT OF LIFE SPAN

We propose that life uses what amounts to a *life cycle strategy* of information collection: It makes a copy of itself with all the relevant information in order to pass that information on. In a manner similar to how an organism's internal information cycles are shared with its neighbors in a cooperative manner, its information is also is shared or passed to the next generation. Often, a same-generation informational or intellectual solution is not possible, and hardware (physical) upgrades become too expensive to make all the intended changes. But another option—leaving the change to its offspring—is available, based on our framework. In other words, death can be seen as a natural part of saving energy and amounts to an evolutionary strategy intended to address a life-form's projection that a new model will be needed under certain sets of circumstances. This also makes sense as it pertains to the survival of younger, more flexible members of a species at the expense of older individuals. Before we move on, we acknowledge that the wording here is difficult. Although we have taken great pains to present our ideas in a way that repeats the subjective nature of life, we want to remind the reader that this and other concepts don't imply that a given organism knows or understands something the way we describe it.

We know from various fields of biology that morphological changes are much easier during development than as a mature organism. This makes perfect sense in that new models may include changes better made during the building process than through trying to retrofit them later. This creates a tremendous savings of energy by producing a new updated model rather than trying to upgrade the older one. Rather, varying lifespans are the natural outcome, given an organism's intentions.

A particular life-form gathers information relevant to the scale of its own particular environment, including demands for changes to meet informational, functional, and structural needs. The filing and sorting (intellectual) portion of the information cycle includes updating the files relevant for passage to the next generation. Included in the knowledge library are plans for a projected optimal structure and a life span. Organisms package information relevant to the formulation of a more fit (updated) version of the same life-form—based on the most up-to-date environmental situation—into the DNA for the next generation.

Because information is gathered and used in representational segments, the relevant segment lengths of various life-forms influence the natural life span of that particular life-form. If new information makes a longer life advantageous, a longer life span will be intended so that action potentials can be passed on.

So the information capacity of a particular life-form is important. A longer life span is useful in collecting a large amount of information, but also as longer temporal sequences, so that the stored representation can be closer to the universal environment. As we have noted, shorter life spans are preferred for more frequent changes or if passing the DNA was the only goal. Of course, this is also dependent on the relevant information available in the environment, including natural survival rate factors such as predators. Basically, if there is no need to update, life would be content with its most up-to-date, built-in life span. If there are a lot of things that need updating or if a given environment becomes more dangerous, lowering the chances to reach its previously normal span, a life-form would tend to adjust its reproduction and life span accordingly. Life is very efficient and will tend to not build the next generation for a long a period if it perceives that predators make that impossible, but perhaps it would plan more offspring or some other strategy to pass its information down. Therefore, the basic influences in adopting a particular life cycle format are signals making the given life span an effective model for carrying out the information cycle on a multigenerational basis. We are not saying here that this type of change can happen quickly. We presume that changes, including changes in the DNA, are based on relatively reliable and tested information, but we do not know all the factors, and this function probably varies depending on the organism.

DEATH AS TRANSITION OF COOPERATIVE STATES

One way to view change is as the disengagement or rearrangement of cooperative function in part of a living form. If there is too much change needed, it results in the death of the organism. Using the concept of life as a hierarchy of cooperative efforts, death can occur when one or more important units no longer carry out the information cycle as part of the cooperative reason they historically came together. In other words, the unique pattern of behavior of the life of an organism due to its centralized information cycle is abandoned or no longer exists. This happens when the reason for enough of the individual units to come together is gone or disrupted in a way in which it can no longer fulfill the conditions of cooperation. In this context, from a subjective viewpoint, a certain internal system or group could actually be healthy but still disintegrates, because the environment that induced its components to cooperate no longer exists. If this happens at a deep layer that can no longer be maintained, death may result. Built-in flexibility has broken down, and core functions have been violated. For example, the loss or damage of certain functional groups or organs may mean that internal environments needed for survival are compromised; disassociations occur and collaborations break apart. If critical functions cannot be maintained, the death of the organism occurs. We have noted that inflexibility, or the inability to repair over time, can cause a breakdown in a system or an organism, and an important part of flexibility is the intelligence part of the information cycle. What we have described here is intelligence at the level of an internal layer reading local signals that can dictate breaking ties that were previously beneficial but are no longer so. Of course, we could also envision local conditions changing enough to directly impair the functioning of local cells or systems. Looking at differing situations from the viewpoint of local life may be helpful in analyzing various health problems.

Looked at in this way, death can be interpreted as a disruption of historical communication patterns. Combining this with what we have just discussed in terms of the life cycle method being an effective way of running the information cycle over generations, we believe that life span, like other functions, is a response to environmental demand as experienced by a given organism. Therefore, internal adjustments can result in a life span that best fits its environmental and internal demands, and an organism will not necessarily spend energy simply for the purpose of living as long as possible.

In part 2 of the book, we attempt to address some issues related to modern environmental information inputs, such as the importance of our intentions and, therefore, where we put our attention, especially during the high-input developmental period. How do we deal with changing technology and its evolutionary new demands on our remodeling mechanism? Can internal remodeling be enhanced to meet the informational demands of today's world? Where do we want to go now that traditional natural selection has a much smaller role? What do we want out of our bodies, and what is our potential?

PART II

INTENDED EVOLUTION IN EVERYDAY LIFE

CHAPTER 12

TECHNOLOGY: THE MODERN HUMAN'S EVOLUTION?

—

WHAT WE COMMONLY THINK OF AS EVOLUTION COULD BE described simply as a change in morphology large enough that an organism is classified as a new form. According to intended evolution, this change is generally based on responses to perceived environmental demands, guided by natural selection. Meeting these external or environmental demands, in turn, leads to internal change where possible. But if life can make changes to some aspect of the environment as a more advantageous or quicker alternative to internal change, it will do so.

STRATEGIC USE OF THE ENVIRONMENT

Interacting with the environment implies more than simply perceiving and reacting to it. Even very basic life-forms

A. Organism selected by nature from a random variety of population

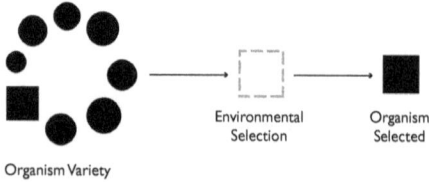

Organism Variety

Environmental
Selection

Organism
Selected

B. Organism changes (adapts) to fit demand

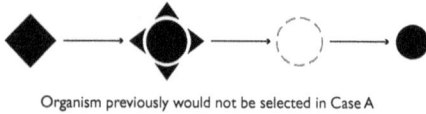

Organism previously would not be selected in Case A

C. Organism changes environmental demand to fit itself

FIGURE 8—THREE SCENARIOS FOR AN ENVIRONMENTAL DEMAND
(a) One organism of many already fits the environmental demand and is therefore selected. (b) An organism changes itself through behavioral changes or remodeling. (c) The organism changes the environment to better suit itself.

use various properties of their effective environment as an extension of themselves, so to speak. Microorganisms, for instance, have developed cilia in order to move; that is, cilia use the surrounding matter as an effective environmental factor to fulfill the situational intention for movement. This is a simple example of internal change in order for a more beneficial interaction with the environment. The organism uses the best resources available, given the characteristics of the liquid, in choosing what to do. Furthermore, this effective environmental factor (viscosity) is very predictable and reliable; the ability to use effective environmental factors relieves some of the environmental pressure and, therefore, some of the need for internal flexibility. In this case, being able to move decreases environmental pressure in that the microorganism can move to change its situation, at least somewhat, rather than just taking what comes to it. The key to using the environment as an extension of internal control is predictability of use. If a life-form can use very predictable aspects of its environment for its own benefit, it will tend to do so, since predictability allows extension of internal control. The foregoing paragraphs are probably recognized by the reader as being analogous to what was discussed in the chapter on information: The reliability and predictability of patterns of information are useful.

TOOLS AS EXTERNAL FLEXIBILITY

Therefore, life not only conforms to fit its effective environment but can also manipulate the environment to reduce its own need for physical conformity. As evolution progresses, more elaborate strategic use of the effective environment or manipulation of that environment can be seen. Animals living under rocks, in caves, or hiding behind things, or even the human adaptation of clothing are strategies that allow organisms to use effective environmental factors in their favor to reduce selective pressures. Essentially, this behavior shows awareness that the organism can manipulate or alter its environment and not just take what comes.

When we see the use of one effective environmental factor against another, it implies that a solution to a problem was found with intelligence and an awareness that something

FIGURE 9—TOOLS ARE EXTERNAL FLEXIBILITY
Evolved physical forms, such as cilia, wings, arms, are just tools, used in the same way as external tools or technology, such as a stone hammer, a wrench, or a computer.

could be done. In this case, internal systems would only receive signals to remodel for the use of the new behavior (e.g., the use of a tool). For example, repeated attempts by birds to use certain foods can lead to beak remodeling over time, but some birds may use a stick as an extension of the beak to reach a previously unreachable food source. The strategic use of the environment can be seen as an extension of physicality into the effective environment, creating an added layer on top of internal stability (homeostasis) through intellectually derived external flexibility. We use the term *external flexibility*, because an organism deals with pressures or challenges from the environment but does so using physical structures out in the external environment. More advanced strategic use of the environment, or manipulation such as the example above, that augments flexibility is normally described as tool use.

An organism can use tools as extensions of itself to augment its interactive capability with the environment without as much internal physical change. Tools are a way of putting one (beneficial) effective environmental factor between the organism and another (challenging) one. We can look at tool use as an extension of an organism's internal self-control, intellectual flexibility, or even remodeling capability, out into its environment. For example, if an animal uses a tool to access beneficial food sources that are not usually accessible, it may save time and energy versus other sources, although it may remodel internally on the basis of manipulating the tool over time.

Time and energy are also saved by changing behavioral patterns to include reliable tools rather than having to resort to internal changes through repeated attempts over long periods of time. Furthermore, reliable tools can also replace previously needed internal functions, such as clothing replacing body hair. Certain tools with high reliability factors in manifesting intentions essentially become almost part of the organism, such as tools for eating food, covering the body, or living in shelters. Considering that energy is the currency of change—and, yet, can be expensive to find and utilize—life always tries to be as efficient as possible in fulfilling its needs. If animals can access a new or needed food source (i.e., obtain energy) by recognizing a pattern (dropping a nut in its shell on the ground or using a rock or stick rather than spending a lot of time pecking or chewing at the shell), it will do so, if it recognizes the possibility. This can either open up new energy sources or make existing ones cheaper.

THE BENEFIT OF MODERN TECHNOLOGY TO FLEXIBILITY

Modern technology works in a similar way as traditional tools: an extension of our internal control into the external environment. Tools imply increased intelligence or intellectual activity: using the same structures and functions to get more done. Of course, this also implies increased physical capacity for that greater intelligence. With the advancement of technology, internal change in humans tends to be directed toward intellectual (brain) change, because our environmental demands are increasingly being dealt with intellectually. Furthermore, as physical activities are replaced by technology, the rest of the body will slowly drop functions when they are deemed unimportant, given a long enough period of evolutionary time.

Dwellings to protect us from the elements or predators, hunting tools, and farming techniques and tools all give us more time for other activities, which means recognizing more and different information patterns. These examples all essentially lead to more stable and reliable core information (basic needs), allowing ever greater centralized perceptive activity, which was discussed in earlier chapters. This is because we do not need to experience so many previously repeated patterns in the form of environmental challenges; they have become externally automated (see chapter 6). Therefore, one aspect of technology is, in effect, a second or higher level of homeostasis that humans are building. Interestingly, like the relationship of the central information cycle in humans to the internal information cycles of the units, technology is built in a similar manner; complex tools such as computers have layers of functions that the end user may not be aware of as he or she uses it.

We know that time is defined by a change in conditions or is simply a concept to explain these changes in conditions (information patterns). Therefore, when we say that we are saving time through a technological tool or have more time because we can now drive instead of walk, we are really saying that we can take in more or different information: dealing with previous challenges has been automated from the viewpoint of the person using the tool. In other words, we have automated one function to allow more attention to be paid to others.

If we walk somewhere every day and it takes an hour to get there and an hour to get back, we experience that walk (information pattern) over and over each time that pattern is repeated. There is no problem with that, and in fact, evolutionarily, it may be beneficial to us if that function needs to be used or perpetuated. However, when we learn a new pattern (e.g., learn to drive a car) and make the trip in fifteen minutes, we can then spend the rest of the time experiencing even more new patterns of information (e.g., getting a graduate degree online). In effect, we can take in more information and have more flexibility as to what information we choose to take in. We potentially increase our knowledge by a large factor and cause our perceptive ability to increase as well. Therefore, even early tool use allowed increased perceptive experiences in the environment, which led to increased knowledge and more diverse and greater perceptions. Previously, we noted that increased perception can also take the form of recognizing more detailed or specialized information, which can be seen in both physical and technological evolution.

Finally, a huge advantage of tools is to allow for energy and resource savings given a certain task. For example, if muscle mass is not needed and is therefore shed by the body because technology does the work for us, the maintenance and energy costs of the physical body are reduced. The small amount of time given this factor is not meant to diminish its importance, but further discussion is beyond the scope of our expertise and purpose here. It is enough to note that modern societies are able to manipulate the environment with little human physical input in a way that would have been unimaginable in the not-too-distant past.

TECHNOLOGY AS A PART OF THE INFORMATION CYCLE

Let us look at technology in terms of the information cycle, starting with perception, as we did before. We can easily see how technology allows massive increases in the available input of information to our perceptive mechanisms without much physical movement or work. Similar to the way the nervous system has evolved over time, the communication system in the modern world is ever more interconnected and efficient. We can see that it actually grows in the same way as human remodeling, based on demand or the needs of the individual nodes—in this case, people, groups, companies, and other entities. When there is perceived demand, infrastructure is planned, evaluated, and built. This is the same as the evolution of life according to intended evolution: A need is perceived, processes are formulated, and structures are built, which results in organisms adapting and evolving. The entire process is similar to the information cycle framework we have proposed: Needs are perceived, (intellectual) decisions are made, and action are taken.

Using technology as a template to reflect back on evolution, we can also see punctuated evolution, when there is sudden, rapid change. Technological evolution seems to explain this very well. There are certain events in which a function is needed (say, processing power) and is slowly expanded using simple incremental changes of past processes, but there is awareness that a new paradigm or method is needed as a bottleneck forms. People research with a goal in mind (projection), and, on the basis of this past information, new patterns and ideas are recognized, as are possible solutions (action potentials build). Suddenly, the right patterns form (paradigms emerge), old information is recontextualized, and plans for new structures and systems are "chosen" and "actions" (building) taken. Once a function is altered as new infrastructure is built or old infrastructure is improved or co-opted, things very rapidly fall into place, related structures are built, and change (evolution) occurs. Therefore, one important bit of information at the right time and place, once it is finally recognized as needed in turning potential to action, can allow apparently large physical changes very quickly, often because the existing processes and structures are simply rearranged to use the newly recognized information. If the change created widespread opportunity for efficiency and benefits, large morphological changes would occur. Typically, old structures are co-opted when possible (new interstate highways, gas lines, electric transmission systems, or fiber-optic routes are typically built over old ones).

In some respects, similar processes occur in technological and biological evolution. According to intended evolution, biological evolution has occurred because a need was recognized and a solution intended by a given organism. Patterns of information were recognized as having potential value for meeting a given need, and a future direction or state was built as an action potential (projection). Over time, as preconditions were met, the organism and its relevant systems became able to actualize it.

THE HOMEOSTATIC NATURE OF MODERN TECHNOLOGY

Just as in the biological model, we can say that the realization for a demand in technological evolution means that there is potential for a mutually beneficial relationship, which is the key to any interaction. As in biology, there is an intrinsic understanding that a perceived mutual benefit necessarily precludes any business or technological relationship; otherwise, it will be short-lived. Of course, the new product inventor wants to benefit in some way, even if that is simply being paid or to fulfill his or her curiosity. Basic research yields a benefit for the researcher by way of the natural desire to recognize new information and put diverse or potential pieces together when possible, which enlarges the societal knowledge base. Curiosity is a well-recognized human manifestation (and it occurs in other higher animals) of intended evolution's proposition of a *basic intention*, and is essentially a drive to recognizing raw information that we spoke about earlier with respect to more simple organisms. Some people look for potential pieces of answers to ideas or problems, which are really the same as unrecognized information. They search based on previous context but also on projections about the future based on an intention to find an answer.

Therefore, in modern technology, which we could call a collective or group version of human evolution, the perception, processing, and choice and action entail broad collaborations. Individual human physiological evolution is driven by the collective or combined information cycles of all interior cells and systems, and in current technological evolution, there are also many units sharing information in a mutually beneficial way. When a new company springs up with a function that satisfies some need, other units will fill in and link to it as a buyer or supplier if that function is beneficial to them. This happens at different levels of order up the hierarchy toward the ultimate demand, driving the formation of the system. As with the biological version, it is not implied that there is an overall understanding of *cooperative group*, but simply that mutual benefit or cooperative efforts are embedded; otherwise, the system will fall apart.

As in the biological example, an action potential can build as new information comes in, options are weighed, cost–benefit analysis is done, and so on. Once a decision is made and potential transforms into action, actual physical activity starts (i.e., infrastructure is built), and the cycle continues as experience is saved as knowledge and fed back to perception. Interestingly, technology is also described as having "life cycles," in which new models replace old worn out ones based on environmental demands and the lack of flexibility.

When we talk about intelligence inherent in evolution, this is not to say that there are not sometimes exceptions or mistakes in which projected benefits do not pan out. If a miscalculation has been made, entities may drop out, and, we could say, the projections or plans were based on unreliable information and end up being co-opted, transformed, or dismantled. As in biology, these are intelligent activities based on the best information at that time. Finally, in technology, similar to in biology, we see situations in which repetition is needed before a proof of concept is accepted, money or energy flows toward an idea, and the idea proliferates into new areas.

INSTITUTIONS AND COMPANIES AS COOPERATIVE EFFORTS

Different companies or institutions, like organisms, can be looked at as intentional cooperative efforts formed in different ways with different functions to deal with external environmental (economic) demands. For example, many different transportation forms get people from point A to point B. How these transport strategies form a mutually beneficial relationship with travelers depends on their inner knowledge, so to speak: how they are set up, who works for them, and what historical expertise they have. Like all intelligent systems, perception is subjective to the internal structure and knowledge—in the case of a company, the combined information cycles of the various people, units, and relationships involved, as well as external infrastructure, patents, processes, and so on. If the external environment changes and internal flexibility to change is inadequate, the company goes out of business—becomes extinct. This can be due to the internal knowledge library being outmoded, to an inability to learn new incoming information, or to many other factors. As in biology, individual units do not necessarily understand an issue or problem in a universal way. Much like intelligently increased muscle mass at the local level doesn't imply a full understanding of the larger muscular system, individual units may not understand or respond accurately to the larger strategic environment. Concrete slabs can be poured locally with little understanding of the *how* and *why* of the overall building plan, analogous to automated systems in biology.

We could call those overseeing the project or investment a centralized information cycle. Furthermore, overall investments or companies also interface, or combine information cycles, as leaders exchange information with the rest of the economy.

At each level along the way, there is only certain information that the next level needs. For example, each worker does not need to tell the company head about the details of a project; the onsite worker addresses the more local details, and the leader focuses on the bigger picture. This is directly analogous to the biological example, and mutual benefit all along the way is the general rule. Thus, if a company or project manager tries to take too much of the benefit, there will be trouble, and a project or relationship falls apart. We can see adjustments in benefit between individual workers, groups, companies, governments, and other entities happening all the time in modern societies.

TECHNOLOGY AS A CONTINUATION OF HUMAN EVOLUTION

Any infrastructure associated with human societies involves sharing effective environmental factors that allow and require cooperation and specialization. Technology creates stability of basic core functions (homeostasis) and external flexibility.

Specialization of labor, like specialization of cells, implies cooperation, increased perceptive abilities for the whole, but increased local stability and overall maintenance costs. Like our individual internal systems, the outer systems are built, used, and maintained cooperatively; they imply mutual benefit and are based on or layered over previous versions of the system. Increased maintenance costs may take the form of taxes, fees, and costs that allow people to forgo doing many basic services. The specialization process means a greater understanding in the specialized areas and increased perceptive abilities of central information cycles through research and innovation.

Technology increases the chances and speed of the actualization of human intentions (i.e., potential turning to action). A big reason for this is the greater amounts of energy that can be brought to bear by the technological layer, allowing one person to do work that would have required hundreds in the past. This has been beneficial to all the levels of the information cycle, from the individual to groups, companies, and governments, in terms of actualizing action potentials.

Historically speaking, we might even say that an early "technology" that enabled external human evolution was language, including writing. In a way, early communication was a crossover between internal and external evolution. Both spoken and written words are physical signals out into the external environment. Although making a tool and showing someone how to use it also fits this description, writing allowed abstract information to be collected, stored, and shared for much more flexible use by wider audiences. This facilitated communication, allowing early groups and then societies to form.

If our physical society, in which so many things are human-made, is external evolution, then its stored information is like DNA, because it contains plans by which things are built. With technology, information transfer is not as limited to physical DNA (i.e., procreation) as it previously was; humans can add to the *collective DNA pool* during their entire lives for the benefit of future generations, or simply a widespread audience in the same lifetime. Of course, modern technological knowledge collection has seen rapid evolution, dwarfing the capacity of the libraries and universities of just a hundred years ago. If we look at information collection (information cycle) as evolution, humans are indeed evolving more rapidly than ever before.

Finally, like the centralized information cycle in biology, the external technology layer also dictates much of modern human's individual incoming sensory perception. This is in line with combined information cycles in biology and with biological evolutionary discussion about perceiving signals on the basis of internal knowledge. Modern humans can be seen as perceiving signals partly on the basis of where in the technological layer we position ourselves. And, according to intended evolution, we will change internally, specialize, and evolve according to signals perceived and intentions formed to deal with them.

CHAPTER 13

REGARDING MODERN HEALTH AND DISEASE

—

COMMON DEFINITIONS OF *HEALTH* ARE USUALLY RELATED TO two basic concepts: (1) the state of complete physical and mental well-being or soundness of the body and the mind and (2) the absence of disease or infirmity. The key terms in the first part of this definition are *well-being* and *soundness*. If we look at the definitions for these terms, we are referred to the condition of the body being healthy and free of disease or damage. *Disease*, the key term of the second part of the definition, is defined as a pathological condition of the body or a disorder of the function of the body, including organs and systems. Essentially, health can be said to be the absence of disease and that disease is the absence of health, which shows that our overall view of *health* is still very vague.

From the perspective of the intended evolution framework, we will attempt to offer our view on some concepts of health. Because of intended evolution's premise that each life-form includes an intentional action potential—information gathering in nature—each cell, functional unit, organ, and so on in an organism can be viewed as an information-gathering and -utilizing node. The effectiveness of each node—essentially, the functioning of its information cycle—can be seen as the health of that unit. This effectiveness, therefore, essentially has three parts: perception, information processing, and choices and actions combined.

Before moving on, we would like to reiterate a point from the beginning of the book: that, of course, these divisions of the intentional process that we call the information cycle are not really separate, but we use this "cycle" for discussion purposes.

PERCEPTIVE ASPECT
OF HEALTH

By *effective perception*, we mean that the node perceives its environment and gathers relevant information as optimally as possible based on evolutionarily relevant situations or needs. For the human as a whole, health professionals monitor perceptive (or sensory) systems such as hearing, vision, tactile functions, for clues about health. When we say "evolutionary relevance" we mean that we must consider the (past) knowledge library of a node when discussing pathology. We cannot expect internal cells or functional units to perform optimally if their knowledge library is compromised in some way (e.g., genetic diseases), but also if the library is in a suboptimal environment, evolutionarily speaking. In the second case, it will, like all life, try to adjust to the new environment, taking time and energy away from other shared information cycles as well as causing them to adjust.

From the standpoint of the intended evolution framework, functions should be in an optimal range based on historical experience. Because the information cycles of all of the systems in the body are shared, even a function that exceeds what is historically needed is not optimal. Unneeded function can be a waste of energy and resources or potentially bring in confusing signals and may take information cycle ability away from other functions. The optimal function of a given information-gathering sensory system is subjective, relative to the organism's evolutionary need. We could say that we are the "latest model" out of a series built to handle *expected* environmental demands.

For example, if an organism feeds only on shrimp, we could say there are optimal wavelengths or behavior patterns the various senses need for shrimp hunting. If the organism constantly notices factors that are irrelevant, this could theoretically be detrimental, noise, or a waste of energy. Another analogy is that humans, when functioning optimally, learn to ignore background noise, in favor of relevant sounds. As we discussed earlier, all of life evolves to recognize beneficial patterns of information. Human senses evolve to pay attention to relevant stimuli, and the healthy sensory systems tune to what is considered beneficial input as well as ignoring or turning off what is seen as irrelevant input. In general, we could say that if the scope of interaction or experience doesn't fit the rest of the organism's information cycle functions, it is out of balance and will function suboptimally.

On the other end of the spectrum, its perception is said to be deficient. This is more what is normally thought of as a problem: when a function can't perform to the needed level.

Deficiency can also occur if the ability is there but is, for some reason, too expensive to maintain versus other needs. This could also be unhealthy in that for something as complex as a human, any function that is deficient could affect the whole.

As for the relationship among the functions of the body, information cycles of cells and systems are shared among the whole organism, as we have mentioned before. Therefore, using our framework, stress or trouble at any level of organization affects the information cycles of related units and the central cycle (entire person). When we speak of *stress*, we are using the term in a broad sense, in the context of the information cycle framework. We can think of stress as a range in which flexibility is challenged or utilized to deal with an environmental demand. We think of the common usage, when there are noticeable stressful feelings, as more when an environmental demand is actually outside the normal built-in flexibility capacity, leading to a temporary impairment of information cycle ability. On a local level, this could even be sent out as pain, which we view as a local stress deemed acute enough to alert the central system. Therefore, local or internal stress affects perceptive abilities for the person as a whole, but also stressful perceptions of the external environment (at the central level) affect internal systems (subjectively).

Therefore, looking to local environmental conditions, shared internal environments, as well as external environmental conditions for possible clues to causes of a given symptom or disease may be helpful.

INTELLIGENT ASPECTS OF HEALTH

The next aspect of the information cycle we have called *information processing* (intelligence) of what is perceived. We expect a given organism to be able to take current experience and process it according to historical experience and to derive an optimal plan of action (intelligent projection), or the formation of an action potential. For example, a core layer of knowledge that functions properly and in a stable manner can be accessed and compared with current perceptions (intelligent processing) in an efficient way, allowing maximized projective abilities. We could say that optimal functioning allows for an accurate and maximized action potential. The capability of an organism to successfully conduct this process is what we are calling *intelligence* for simplicity's sake, although we could think of this as *mental capacity*, or mental health in humans.

The more intelligent an organism is, the more relevant information it can process. So if an organism's intelligence is incapable of smoothly handling the information available in its effective environment, we can say it is unhealthy. Looked at in this way, according to intended evolution's framework of combined information cycles, a declining mental capacity is not simply a brain or nervous system problem. Because the information cycles of all the systems are combined, perceptive and intellectual problems can be due to pathologies in other parts of the body. One aspect of our framework is that internal disruptions may cause a situation that takes intellectual ability from the central system to use locally until the situation is rectified. We can easily see with this model, for example, how chronic problems during the aging process would always affect perceptive ability regardless of whether there is a "brain problem" or not. An extreme example is that low blood pressure from any etiology can cause a decrease of perceptive and intellectual functions. Essentially, any trouble anywhere affects mental functioning to some extent. In our framework, this is an example of the *soundness of the mind* mentioned in typical definitions of heath, and of course, medicine today uses many diagnostic tools to test perceptual and cognitive capacity.

PHYSICAL ASPECTS OF HEALTH

The final part of the information cycle is that of choices and actions. Once an action potential is activated, can the organism execute optimal and efficient action? This may be thought of as the more physical part of health: Is the body capable of acting on the information commonly found in its effective environment? Of course simply being alive indicates that a given individual has the latest "hardware" to sustain life according to the most current environmental data collected, to the degree possible.

This portion of the information cycle is more about execution by the inner, built-in layers, which are relatively stable and predictable, as we have noted before. This stability also holds back the softer yet initiating intellectual portion of the information cycle, in that physical movement and change takes time. Generally, evolutionarily inner layers of the body are typically more specialized. This means that those units normally use more narrow perceptive and intellectual processing ability and tend to include more automated features from the whole-organism perspective. Therefore, local units having to do with physical action are generally more fixed and automated to react to a given input from the central cycle, or mind. This is not to say that all signals from the central cycle are equal, and what types of signals can affect which internal layers and why is an area we believe should be investigated.

If a local internal environmental change or any other situation causes local life to ignore or pay less attention to its original plan of communication, this affects physical functioning. This relates to anything that deals with changes that need to be made or any unit that needs to return to baseline cooperative functioning. This could be anything from a muscle returning to resting state to a rebalanced endocrine system. Clearly, situations in which any system cannot return to baseline could be very disruptive to related systems and the person as a whole.

THE HEALTH OF CORE FUNCTIONS

Through evolution, related cells or groups that were once separate came together in a cooperative manner to form functional units or systems with varying degrees of specialization and what we have called *automation* and *flexibility*. In general, functions that are more highly automated, such as instinct, the heartbeat, or other critical activities, are what we think of as part of the physical core. Core attributes are very expensive to change, requiring long periods, or generational time frames. Smaller and generally evolutionarily more recent physicality and function (softer attributes) are easier to change.

Some portions of a complex life-form's energy-gathering mechanisms are an example of what we think of as core functions, and are typically somewhat automated internally. We say "somewhat" because some flexibility exists—dietary changes, for example. With energy gathering automated, more attention can be focused by an organism's centralized intentions and intelligence. By this, we mean that to the extent that digestive function is automated from the viewpoint of the central awareness, it allows greater perception of signals from the external environment. This does not mean that the digestive system cannot remodel but simply that historically parts of the system relied on stable local conditions (food sources) being supplied by the organism as a whole.

With energy and intention, an organism can make changes as well as affect its own future while still ultimately being based on optimal functioning of core systems built throughout evolutionary history. If an organism needs to change energy (food, here) sources and so isn't delivering it optimally where needed, the central perceptive and intellectual phases won't operate optimally, as the central information cycle is weakened by internal, or local perceptive needs to switch to new input. In this case, there would need to be a rebalancing period, assuming there is enough flexibility to make the conversion. Conversely, when food sources must change, internal needs are signaled to the central mind and used by intentions at the organism level to find alternatives. Other "core" functions may be more or less automated and have varying amounts of flexibility. The heart, for example, can remodel to some extent, and yet its ability to adjust and keep functioning, given modern internal environments, is limited, as is evidenced by the rise in heart disease.

As we have said, we view either excess or a lack of core function demand according to historical norms to be suboptimal, meaning that a function can be overused or underused. This is because both lead to the use of internal flexibility to rebalance the situation. Too much demand to change stresses the remodeling capability, but so does a lack of demand, which can also lead to the remodeling (dismantling or atrophy) of a system. Core functions are less flexible and, therefore, are harder, take longer, and are more expensive to change. Therefore, when any core function dips under the optimal information cycle range in any portion, we can call it *unhealthy* or *diseased*.

COOPERATIVE RELATIONSHIPS IN RELATION TO DISEASE STATES

We often focus on the symptomatic treatment of our bodies and tend to treat local systems directly rather than treating the process that led to a disease. The highly automated nature of many core functions lend themselves to this method, and many advances in medical treatments are based on the cause-and-effect nature of these systems. Using the framework presented here, we look at these systems as once being more independent and as having formed intelligently through evolution and having degrees of automation (cause and effect) versus intelligent activity.

According to the intended evolution framework, human physiology can be looked at as a description of the evolution of intelligent processes and collaborations. Physiological processes are being done intentionally by structures that evolved intelligently throughout history and also form intelligently during development. The body tries to optimize its functioning, given current conditions, and therefore, internal trouble or disease could also have cooperative components if the original conditions of cooperation between units is changed or challenged.

According to intended evolution, multiple life-forms or cell types came together only when there was a mutually beneficial situation (i.e., cooperation) or, we could say, there was or is a *shared demand* from the common environment that was not as efficiently satisfied by any one life-form alone. Together, there was more information or more usable energy to create an intended action. In this way, we could say that multicellular life is about processes to deal with common external demands.

Therefore, for multiple life-forms to come together, there must also be a common purpose, so to speak. Because evolution developed this way, this is true for all tiers of organization within our body as well. Different functional groups or systems, such as glands, digestion, immunity, respiration, and so on, all evolved on the basis of shared functional demand and cooperation for mutual benefit.

When any of these organizations starts to act in a way contrary to the intention that brought them together in the first place, that is unhealthy behavior from the combined or whole-organism standpoint. If a system or organization exhibits inability to fulfill the demand the cooperation is based on, we can say that the group or system is unhealthy. Also, if the demand for that group's function is too low (or no longer exists), the group could appear unhealthy, or start to atrophy.

Finding what factors the body is reacting to that lead to the undesired states can yield ideas for correction. But viewing what we call undesired functioning as intelligent and possibly healthy functioning given current local conditions may yield a new perspective on some disease states.

Using this framework, can we form a comprehensive treatment to augment current therapies by inducing the body to turn pathological actions back to their original function? For example, can we create the right local environmental pressure to induce stronger intercellular communication and function? This may result in a correction of functional disease and may increase the effectiveness of current treatments for functional disorders (see chapter 16). Capitalizing on the body's flexibility and remodeling capabilities to augment a treatment rather than trying to attack each problem directly may be beneficial. Obviously, this method is more preventative in design and may not be the best solution in

cases of injury, many of which require immediate intervention from outside of the body. According to the intended
evolution framework, recognizing problems as early as possible is key, because much pathology can be seen as a form
of intentional remodeling, which is energetically expensive
and more costly to reverse the further it progresses. Furthermore, both processes take away from flexibility and ability
elsewhere in the body. The potential benefits of a therapeu-
tic approach including the intended evolution framework
are many—methods based on this theory could be used to
augment current recovery process as well as prevention (see
chapter 16).

Often, with certain chronic conditions, the current
model of disease management is to wait for the condition to
worsen before serious treatment can begin (e.g., hypothy-
roidism). But by passively waiting for conditions to deteriorate rather than being proactive, opportunities for correction
are lost, and eventual reversal or cure becomes more and
more expensive as far as the body is concerned. On the basis
of the life-cycle method of determining a life span, it is not
good for health and longevity to let problems persist, thus
signaling to the body that it is in decline.

POTENTIAL HEALTH BENEFITS
BASED ON INTENDED EVOLUTION

The modern world brings new challenges to the human
body in terms of new functional demands, as well as rapidly
increasing and widely varying environmental signals. Under

certain circumstances, we could say that these can be what are commonly called *stress factors*. We think of the common usage of stressful feelings as when an environmental demand is actually outside the built-in flexibility capacity, leading to a temporary impairment of information cycle ability. If any portion of the information cycle is overloaded and gets backed up, we can call it stress. Any unit of the body, being intelligent, can be stressed in this way. For example, evolutionarily unrecognizable or new factors in the diet or in the environment can directly stress digestion or other systems they come into contact with.

Many challenges need to be met with the understanding that our body's systems are intelligent and, at some level, are doing what they do on purpose. What effective environmental factors are these systems reading when they do what appears to us to be something inappropriate? How do we maximize homeostatic balance and remodel flexibility to best meet modern challenges? How can we control our effective environment, and what effective environmental factors can we interact with for our benefit rather than in a way that may be to our detriment? What do we intend to do in the future, and is current incoming information congruent with these intentions?

Humans can change their environment by changing exposure to many signals from it, as well as by using their ability to formulate their own virtual environments. They can do this at almost any time, not only to reduce stress but also to proactively plan ahead so that stress does not arise. With the right plan, virtual environments can induce changes in body function to help deal with current or pending problems. By planning a health care offense—an intentional strategy—based on using body functions to help correct or prevent predictable future events, we may be able to delay or even eliminate the need for defensive medical strategies in some cases. This idea is not new; for example, most people recognize that exercise and diet are the basis of good health. These actions can help fend off future trouble, but we may be able to augment these basic activities for the modern world. We should look ahead, project what future troubles may be, and avoid them or augment strategies that counteract them. With proactive planning the body automatically tends to avoid trouble, just as it tends to hit a better golf shot if you've pictured the perfect one in your head. If you have planned ahead and expect what is coming, you automatically avoid what could have been stressful, for example.

All external demand could be seen as a stress for our body to handle. But the situation in which the body has had time to prepare is relatively less stressful. This is similar to the principle of immunization, which readies the body for future events. The body is at its most stressed when it is trying to deal with a large amount of new and highly important data that must be processed in short time frames, both mentally and physically.

Besides physical trauma and genetic disorders, which are often outside of our control, stress is a major root cause of our health issues. Stress to the body could come from the mind, from physical activity, or from the environment, in the form of foreign substances. We could design situational demands for the body that are aimed at strengthening and therefore preparing specific systems that we know will likely to be stressed in the future.

HABITS AS AUTOMATED PROCESSES

A counterintuitive example of internal intelligence at work in response to perceived environmental demands can be found in bad habits. However, habits aren't necessarily good or bad but can, instead, be looked at as a tool used by the body. Habits are formed by automation of information cycle processing of a repeated situation. Repetition trains a given system or functional group to act in a certain way quickly and without much processing. However, this can be changed over time essentially based on how long those responses took to automate, and by the same mechanism that was used to form them. With proper environmental perceptions (e.g., virtual) and new responses, many habits can be changed, as evidenced by methods used in psychology today. Can we use the information cycle to help stimulate a desired response or to induce the body to correct itself? Perhaps we can also create comprehensive packages to formulate corrective simulated environmental demands, both virtual and with tools, for specific diseases. These methods might be employed as prevention, treatment, or to augment existing medical treatments.

CHAPTER 14

REGARDING MODERN SPORTS AND FITNESS

—

SPORTS AND FITNESS HAVE BEEN A PART OF HUMAN ACTIVI-ties for thousands of years. They have recently been studied extensively, and they are important in most cultures. There are many ideas as to what it is to be fit, but *fitness* is generally defined as being in a healthy physical condition. Fitness has also become closely associated with sports, the definition of which varies but usually involves competition to display physical excellence.

Although we are not using it here in the strict biological sense, intended evolution views fitness in a broader context, loosely as excellence in efficiency with information cycle tasks. Looking at sports and fitness, most of the activities are more or less a combination of these factors: perception, intelligence, knowledge, and physical execution. The term *fitness* is usually associated with physical health or strength; for our present purposes, it is the physical expression of control and flexibility of the information cycle. Furthermore, we don't want to confuse our usage in this section with the biological or evolutionary usage of the term. The more an individual is capable of manipulating his- or herself and his or her sports environment, the higher is the chance that the individual's intentions will be fulfilled.

FITNESS AND ENVIRONMENTAL DEMANDS

In the world of sports, an athlete creates an external demand for him- or herself when training by simulating a given competition. The training space or competition arena is an environment to interact with in order to train the mind and remodel the body for the given demand. This pushes the body to respond in a way that improves certain existing functions and their efficiency. By challenging a function or system's flexibility to its normal limit (but not so far beyond as to produce stress), we can alert the system that the full function is needed.

Functions or skills such as movement, speed, weight-bearing power, quick judgment in activity, cooperation between individuals, and so on are all tools developed in the evolutionary process to deal with the historical demands of a given effective environment by means of our intelligence. Whatever the demand is from the environment, our mind and body will try to respond intelligently to it. However, if the demand does not represent a substantial pattern of change, the body will not recognize the need to change in a substantial manner.

When looking at the effectiveness of any sports or fitness program, one common point is that of repetition in response

REGARDING MODERN SPORTS AND FITNESS

to a given demand. As was pointed out earlier in the book, this is essentially how life works: Through repeated perception and action, knowledge or memories are saved. In the case of sports and fitness, the term *muscle memory* is commonly used to describe an example of knowledge intelligently acquired through repeated activity. Under the constant demand of one type of activity, physical changes will occur to improve efficiency in that area. When we look at marathon runners or weight lifters, for example, we clearly see a change in body shape corresponding to that particular demand on the body. As was discussed in chapter 13, when external demands are made on the body, the more intelligent the reaction is, the smoother the information cycle process is, meaning that there is relatively more flexibility and less stress on the body. In other words, the body tries to get the job done in the most efficient way, using existing built-in flexibility of functions that make the most sense for a given demand.

Therefore, according to the intended evolution framework, when the mind (central information cycle) is engaged in an activity or exercise, the response to the demand should be more effective. This is because local (say, a given muscle) and central perceptive input is consistent; the mind is focused on lifting the weight while the muscles lift it. This is opposed to some modern exercise routines in which the participant is jogging on a treadmill while listening to the news or watching television. According to the intended evolution framework, this is clearly not as effective as paying specific attention to what you are doing. Furthermore, having clear, intentional goals about the outcome of exercise or training is very important. Intending to reach goals and formulating plans to work toward those goals is important to health and longevity, like any other endeavor. Keeping a goal in mind often enough, or, we could say, holding an action potential, increases the chances that behaviors will fall in line with that goal. Automatically, patterns of information consistent with the goal will begin to be recognized preferentially to those that are detrimental to the goal.

Modern sports psychology techniques are good examples of how and why this type of activity, which can be called *mind–body medicine* or *fitness*, can be effective (see chapter 16). An athlete has specific intentions and visualizes a plan (goal) before each action. These include both mental and physical routines for each situational intention. From the standpoint of the intended evolution framework, any fitness plan should formulate carefully what is demanded of the body and should include both visualization and physical activities. By visualizing an outcome or outcomes, we are internalizing information patterns that are used for creating action potentials. We could call this formulated demand a *virtual fitness environment* that the body interfaces with to achieve the desired result. When the intention is focused or clear, there is higher level of synchronization of execution throughout the tiers of organization of the body.

MIND–BODY FITNESS

Mind–body fitness programs, in which the mind and body are both engaged in a given activity, can be created as described above. They can be based on situational intentions that are needed to induce internal changes and remodel in various ways. Through repeated or held intentional goals and direction for health, for example, the body will tend to naturally avoid any behavior or condition that would conflict with the goals. Of course, simply intending to do something has limitations, and we are speaking here of tendencies based on repetition, which are also much more effective when they include a physical context or activity. We believe this is because information cycles are not separate, and concordant experiences at multiple levels form more integrated memories, although we do not know why or how memories are stored at the various levels.

That said, internal intelligence automatically responds to scenarios run in the virtual environment (visualizations), and we propose that visualization and intentions will magnify this message and will augment the effectiveness of a physical program. As part of this process, fitness programs should also maximize the body's flexibility to deal with today's world in terms of our evolutionary past. We also need to plan now for challenges that the body will face in the future so that the stress on the evolutionary process and on ourselves is lessened.

REGARDING SOME CURRENT FITNESS PRACTICES

Clearly, physical activities are important in staying healthy, using the model that physical activity is a link of the information cycle and therefore helpful for stress relief in that it can tend to subdue mental activity as a side effect. Today's fitness model pertains mostly to creating demands that increase heart rate and overall metabolic rates (i.e., to burn calories to try to burn fat) by placing situational demand on various muscle groups. A side effect is the demand for larger muscles, and in some cases, this is also a goal. But are the various strategies designed to place a demand on the body to burn the maximum number of calories in line with an overall goal of fitness and health? In many ways, the current calorie-burn model is a very limited view of the overall human condition, especially in the context of the entire life cycle and other aspects of the intended evolution framework. For example, does the current model of raising the metabolic rate to burn calories create a demand on the body that helps increase its response flexibility to create a more efficient information cycle? Do the results, including the remodeling they induce, increase health and longevity?

OPTIMIZING FUNCTIONAL LIFE SPAN

In the natural world, longevity is often inversely correlated with high metabolism rates as well as with high caloric intake, so further study on the human example is warranted, in our view. For example, current fitness programs often entail a significant amount of stress on the body, especially the current movement toward the mega-workout and ultra-high demands of professional sports. A high level of remodeling demand, which we believe could contribute to shortening a person's life span if too extreme, should lead us to question the attitude that more is always better. The more changes that are required of the body, the sooner it may conclude that the time to give way to a new model (i.e., that it should end its life cycle) is at hand. There is also the localized wear and tear that comes with excessive high-impact activities, which can inhibit needed functions going forward. The common practice in some professional sports of extreme remodeling resulting from very high demands on the physical body may have led to poor health relative to other athletes in the long run.

Of course, we also know that if they are left unused, the body's functions will deteriorate, and it is understandable that calorie burning has become a good short-term answer to a lack of activity or excessive caloric intake. If a person does not walk, all the functions associated with the activity of walking will decrease, leading to health imbalances. Too little activity manifests in many possible physiological symptoms. Humans are clearly meant to use their movement functions, and not doing so affects many systems. Today we often see both of these problems at once, in that lack of activity followed by a flurry of it in order to make up for past inactivity essentially sends contrary signals that functions aren't needed followed by needing them on a near-emergency basis. In this way, either too much or too little demand on our systems leads to stressing our internal flexibility capacity as the body remodels to fit the level of activity. Today's lifestyle of more sedentary work conditions has created a large demand for efficient activities that satisfy the body's need to use evolutionary functions but in a relatively short time.

So to truly have a fit life—that is, to have a good level of proficiency in core life functions for an optimal life span—one must achieve a balance between performance and cost in remodeling stress on the body system.

To achieve good overall functional efficiency during the largest percentage of a life span, the pursuit of purely short-term performance-based goals should be avoided in favor of long-term ones. Making short-term speed, power, endurance, and so on the end goal may force the body to strive toward those changes at a cost of functional years of the total life span. On the other hand, planning on using these same functions throughout one's intended life span leads to a different usage of them but also one that keeps good function.

From the point of view of the intended evolution framework, a functional information cycle across a full life span is the point of fitness. We can attempt to achieve this by not focusing on short-term performance-based goals as the core demand of a program but, instead, focusing on extending the *functional* time of the life span with longer-term goals. We should try and do this by designing exercises that place demands on the information cycle to increase information gathering without undue stress. If we can then add in shorter term, or situational demands, such as speed and power and other functions, in the context of serving the longer information-gathering cycle, rather than as ends in themselves, long-term benefits accrue over time. This can guide the body away from something that is not optimal, or, conversely, to do a bit more because there is an overriding goal linked to the activity. So the intention is not to run fast but, rather, to move at a speed that is helpful for achieving the best functional life span according to each individual's goal, which, of course, may include speed at various times in one's life.

INTENDED FITNESS

Every life begins with the knowledge of all the evolutionary history up to that point as expressed in its form. The body of a person is the structural and functional reflection of his or her particular branch of life and its intelligent, historical, interaction with its environment. However, we see the basic processes of the information cycle embedded in every structure or system via their behaviors within the body. Therefore, physical structures also have the mental component. What kind of change we want our bodies to create is determined by intelligence reacting to environmental information, or demands. While energy is the currency of change, the intelligent interface with environmental information directs it.

What demands can or should be placed on the body and which ones are best avoided? What environmental demand created the need for a person to do many popular modern fitness activities, and are they in line with our evolutionary past?

ACHIEVING BALANCE IN FITNESS

As we have said, if you do not use the functional units at various levels of organization within the body, that inactivity, in itself, becomes a demand for change since it is different

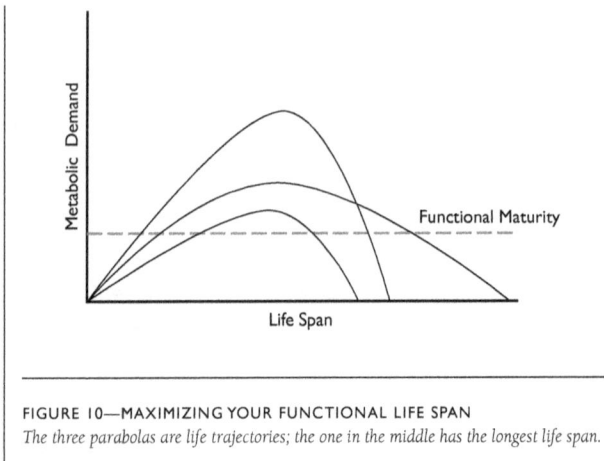

FIGURE 10—MAXIMIZING YOUR FUNCTIONAL LIFE SPAN
The three parabolas are life trajectories; the one in the middle has the longest life span.

from the environmental demands that the body is intelligently evolved to interface with. This leads to a deterioration of the organization of the unused unit, even for core systems such as the skeleton, given enough time. How to find that balance of functional stimulation while managing stress and undue wear is what we should keep in mind when talking about fitness.

According to figure 10, physical demand should follow a curve in which most of the activity falls above the functional maturity line (the dotted line). As far as fitness is concerned, we should manage our exercise (simulated external demands) in a way that is within the best trajectory for a maximum intended life span. Looking at the figure, we can see that a remodeling demand that is too high makes the life line too steep; it will crash earlier. On the other hand, too little activity would result in a slope that is too flat, without enough metabolic demand, which will lead to a loss of function.

We should try to achieve a balance of performance over the life span by staying within stress or remodeling levels that do not lead to early wear and the loss of function. Using the figure, we should try to keep remodeling (i.e., the use of flexibility) to a level that does not create too steep of a curve and, therefore, a shorter life span.

If a person has a clear long-term plan in mind, the outside demand will tend to be seen as an opportunity for the body to realign into something already planned for or a situation best avoided. The results will save energy and will be a lot less stressful than a forced response to a short-term need to lose weight fast, for example. When designing a fitness program, the key is to create workouts that fit into a long-term plan.

CHAPTER 15

LONGEVITY AND LIFE SPAN EXTENSION

—

WE THINK OF LONGEVITY AS THE MAXIMIZATION OF THE FUNC-tional life span—that is, a long life that retains a relatively high level of function. Therefore, it is not necessarily about the longest life span but the best life span for a given body. When speaking about fitness, longevity, or health in a fundamental way, it is not enough to look to the static condition of a cell, an organ, or the DNA. We believe these things should be viewed in relation to interrelated cooperative systems including the expected life span of the cooperative whole. For example, certain organ conditions might mean different things, depending on related conditions or even at different ages.

Even though there are clearly limitations to what the body can and will do, we believe it is important to explore the role of our mind as it relates to human potential in general, as well as how it pertains to longevity. This can also help us use our intention to maximize our functional life span in an effective way. In doing so we put problems such as health and fitness in a larger context, since if one were to increase the functional life span, it also implies increasing the level of current health and fitness.

CONTRIBUTING FACTORS OF A LIFE SPAN

The natural life span of an organism can be looked at as an information-gathering cycle period for that generation of the organism's species. As was described earlier, a central cycle consists of or is made up of the cycles of internal units or systems.

The body's systems constantly use their homeostatic and remodeling capabilities to adjust to its environment: External demand for change destabilizes homeostasis after which the body rebalances. Over a lifetime, the body loses flexibility, especially if demands constantly keep the system out of balance—for example, from chronic stress. When an organism is unable—or we say, doesn't have the flexability capability— to meet new demands, pathologies can develop and, if they persist, can eventually lead to death.

Centralized information processing and intellectual capabilities are possible because of internal automation or homeostatic systems, which create the low-cost core stability needed for the central cycle to operate optimally. Therefore, demands on the perceptive and intellectual functions can be held back by the limitations of the physical structure of the body, which is more time-consuming and labor-intensive to change. This works both ways; excess intellectual activity (stress) can compromise internal units and functions, and compromised internal units mean a lessening of the central intellectual ability. The information cycle needs to run at a balanced rate to achieve maximum efficiency in forming and executing action potentials. As we have said, demanding too much wears the body out; too little demand, and it withers.

As we saw earlier, information collection during the life span will depend on the life-form and its life cycle strategy. The likelihood of unnatural death influences how much information exposure a life span is likely to have before needing to save that information in the form of reproduction. If an organism perceives a hostile environment, the average expected life span of that life-form will be shortened because of a high "unnatural" death rate, and the reproduction cycle will be purposely adjusted. Therefore, under these conditions it would not be efficient to adapt a long life cycle, because the chance of reaching that length is minimal and the chance of passing on the data gathered during the life span is greatly at risk.

When looked at using the intended evolution framework, if life were truly driven entirely by the sole purpose of passing down genetic material rather than collecting it as well, there would be minimal need for anything but a short life span. The most effective way to protect genetic material and pass on random changes would seem to be with progressive rounds of replication for selection to work on. So life's pattern of evolution would lead to as short a life span as possible with as little developmental and as few structural requirements as possible.

According to intended evolution, there is a tendency of the internal drive to expand and increase the information cycle capacity, which also necessitates self-preservation behavior in order to protect its knowledge and information-gathering capabilities. Longer life spans and greater developmental periods (time of great information collection) also make sense using the intended evolution framework.

Understanding more about why we can intend to live to a certain age with a given body allows proactive formulation of realistic intentions and plans rather than just to act defensively to try to stop various aging processes. As we have said, realistic planning for the future automatically reduces stress, in that surprising and stressful events are less likely to occur.

Throughout this book, we have noted that the evolution of life to more and more complex forms resulted in an ever-increasing perceptive ability as measured from an organism's current position in space and time. We believe longer life spans are a natural result of this basic intention to accumulate information, if it is combined with an environment that induces an organism to plan for distant events. As we increase our life spans, we increase the scope of our information cycle abilities, including projections farther in time and distance, allowing us to increase our control over our own futures. Therefore, this natural quest for knowledge requires expanded information-gathering capability, which is available by increased functional life spans in the future.

LIFE SPAN EXTENSION

This question is not just how we can live longer by avoiding disease; it becomes how can we intelligently design a future in which our life span is naturally increasing. When we talk about life span extension, we mean lengthening the natural life span not only for this generation but also for future generations.

Some factors that influence life span include the built-in projected life span derived from past experience, the information cycle capacity of the current organism (including flexibility or remodeling capability), and the amount and type of effective environmental factors an organism encounters. Of course, these are interrelated, but we think a couple of strategies are worth discussing, given the overall goal of an optimal functional life span.

If we look at the projected current life span, we could formulate a strategy to limit one's external environment so that there are very few new effective environmental factors. In other words, try to keep the environment stable and match it to that person's evolutionary past. This makes sense, although it seems to conflict somewhat with the growth of the information cycle we see through evolution of complex organisms. Regularity is the key element here; with low information cycle growth and a life of very few surprises, necessary remodeling is kept to a minimum and one can use energy for a longer, relatively change-free life. This strategy would use all of the evolutionarily important available functions in the body in a consistent, balanced, manner while minimizing stress. We could call this "riding the line of optimal use" of a given body with not too much function demand so as to limit stress of internal flexibility, and yet enough demand on the important functions that the body has evolved to use. We say use all the important functions, because underuse of any may lead to deterioration of that and interrelated functions.

If a person were able to avoid life-threatening situations or disease, a long life span with relative good function is highly likely. Many life-forms whose effective environments are relatively stable—such as certain trees, and low-activity animals like turtles, corals, or some lobsters—have this kind of life cycle that fulfills the evolutionary history of that given organism. This type of activity is also seen in humans living simple lives in certain areas of the world.

Next, if we look at a given information cycle capacity, we could say that a longevity strategy of active engagement with new environment factors and maximization of the information cycle makes sense. This lifestyle usually comes with a prerequisite that basic physical survival is achieved with relative ease because of a previously learned capability to handle the environment. Without this prerequisite, the information cycle capability will be used for short-term challenges to survival. This capability is evident in the ability to meet challenges intellectually rather than a way of living based on physical remodeling, which is energetically expensive and not always conducive to longevity. With this approach, the person can handle most of the physical challenges of the environment through "soft remodeling": using intelligence and current function rather than physical restructuring to deal with external demands.

In humans, large amounts of new information can be processed, and this strategy is predicated on including long-term intentions and projections in the information-gathering process. This creates situational intentions to achieve what naturally leads to an internal demand for a longer life span. Connecting to longer-term information signals and demands rather than many short-term ones basically creates a central demand such that the current physical life span length is limiting the efficiency of the information cycle. By making projections based on demand for more long-term information patterns to be gathered than is allowed by the current expected life span, the body naturally works to increase it. Short-term and more expensive demands then tend to be ignored or avoided in favor of more favorable ones over time. We could also say in general terms that by making plans for a very long life and holding that intention (action potential) to the extent possible, some detrimental activities are naturally avoided or dropped.

We believe the consistent pursuit of a combination of both strategies would yield an increase in the natural life span over time, including genetic changes that are needed. In the first case, the need for a longer life span is created through the security of survival and a low level of new environmental demands to deal with. Because there is very little change and the available functions can handle the demands of that environment consistently, it would be efficient to not limit life span: to let the already functional organism that is stable in its interactions go on as long as possible. In the second case, there are a lot of long-term information patterns coming in, which implies that more time will be necessary to recognize the patterns of that environment; a life-form's internal units will respond by actively extending their life span, because that extended time is needed, or demanded, by the central cycle for further information cycle expansion. Furthermore, in modern human society, technology can reduce demand on physical flexibility—or, we could say, it can reduce challenges to our core (e.g., homeostatic) functions. While demand for information cycle expansion generally leads to increased life span, it is technically the need for long-term information, knowledge, factors, or goals that will drive a longer life span.

ADDITIONAL FACTORS REGARDING LIFE SPANS

In any particular case or lifestyle, there is an important factor that will shorten the life span: external demand for too many physical changes that challenge the available flexibility such that they reach the evolutionarily core structural level. If this is the case, any organism will tend to hold action potentials, or intentions, to shorten its life span as a reflection of demands to make changes to the current deep structures. Deep structural changes that challenge flexibility too severely are expensive and will tend to lead to a shorter life span. All the internal systems and units are intelligent and update their intentions on the basis of current input. If the demand on flexibility is greater than the historical norms such that system repair is compromised, this may lead to a shortening of a given unit's projected life span. This information may be shared throughout the system, creating action potentials leading to changes in cooperative agreements or function currently or in future generations.

This is why we believe that we should not make massive amounts of physical remodeling demands, which will indicate to the internal systems that they need to frontload their use of flexibility in favor of lasting longer (see chapter 14). This is a concern in fitness today, because we believe that it has become too focused on current performance. Some activities may bypass our intelligent and functional layers of flexibility meant to deal with environmental demands in order to challenge inner layers of flexibility directly and repeatedly. Of course, we may benefit from the changes in the short term, and it may even allow more offspring—but potentially at the price of a shorter life if sustained for long periods of time.

The human body is very flexible and, with the help of tools or technology systems, can handle extreme external demands. Even our physiological remodeling functions within the human body have great capabilities. Our body can adapt to many demands in a single lifetime, but this ability—or *flexibility*, as we have been calling it—is also used to repair problems, which compete with remodeling for such things as sports performance.

Human intentional projections can influence much of what is demanded of our body (see chapter 16). We can form a new path, along which we can achieve both stability (limited effective environments) and stimulation (active, intelligent engagement and knowledge collection) in order for us to both maximize what life span we are given and push to increase it as we evolve, going forward. Ideally, we can incorporate regularity into life physically by having a very stable life pattern (use the same base physical environment) based on our evolutionary history. By eating and sleeping at the same time, by eating similar food groups (that we have evolved to eat), and by doing exercises that appease existing physiological functions we can reduce demands on the bodies' core functions. An example of a core function often challenged today is trying to adjust to sleeping during the day and working at night. At the same time, we can engage in new informational content, which would encourage a longer life span while reducing excessive short-term demands (reducing stress). Using intelligent planning or goals to ease the stress of adaptation and remodeling and avoiding quick and surprising changes is simple and common sense. We could let technology handle short-term hourly or daily stress (rather than creating it) while having long-term intentions and longer effective environmental factors (planning years and decades ahead). If the information patterns that we are dealing with are getting longer and longer, there will be a situational demand for the life span to be longer as well.

CHAPTER 16

MIND–BODY MEDICINE

—

THE CENTRAL INFORMATION CYCLE OF AN ORGANISM, AS was described earlier, is a cooperative combination of perception that represents the group as one unit (the organism). For the convenience of discussion in this chapter, we will loosely call it an organism's *mind*.

With respect to health, including the role of the mind is often referred to as *mind–body medicine*. The term can be loosely used to cover modalities in which the mind is used to affect the physical body or, more generally, using the human mind to affect health. Although the efficacy of mind–body medicine is not necessarily generally accepted scientifically, there is enough evidence that many well-respected universities and medical institutions have dedicated resources to this emerging field.

We believe that mind–body medicine is in its infancy and, over time, will become a very important part of the established health care community. As can be seen in part 1 of this book, according to the theory of intended evolution, the mind and the physical body are not separate entities. We can even look at the physical aspects of the body loosely as tools that have been developed intentionally through evolution for the mind to fulfill its intentions. Even inside single cells, this idea holds, in that the organelles, scaffolding, and even the DNA are constantly being manipulated for specific purposes, not unlike our use of tools.

PERCEPTION AND THE MIND

As was described in chapter 9, the subjective perceptive abilities of different organisms reflect their interface with the external environment throughout their evolutionary history. This is also true of the physical human: It is a reflection of its information cycle function throughout its evolution up to this point. We also described a given organism's mind as reflecting a shared portion of the inner units' perceptive ability, whose purpose is to enable the internal life to interface with the environment as a single unit. This is also true of internal functional units; they also share information cycles to act cooperatively to fulfill a certain function demanded of them. Or, we could say, their combined information cycles act as one with respect to that function, which is a shared demand. As complexity increased in evolution and became more specialized, communication pathways that were often used demanded and therefore developed more reliability— or became hardwired—by members who specialized in communication in cooperation with others in a group. Although large portions of the body could be said to have developed varying levels of low-cost, stable, and reliable automation, these are not separate from other less automated and more flexible units or functions. All levels are in constant communication in their own, subjective ways.

Therefore, each cell in the human body can be said to have its own mind, larger functional groups of cells share information cycles of the individuals and therefore also have a mind, and so on. Each level of hierarchy or tier acts as a unit, but each tier is also part of the next tier and so shares perceptive abilities with relevant others and the central mind. Furthermore, what the organism-level mind perceives translates back to the inner tiers, including the cellular level, through the tiered layers of intelligence. By viewing the human body in this way, all of its processes and structures are the result of historical layers of life that have been added as well as updated by new layers during evolution. Or, we could say, that these processes and structures have all developed subjectively and purposely on the basis of mutually beneficial interactions at that time in the organism's evolutionary history and then updated through time as new layers of information update the old.

These internal tiers adjust on the basis of the demands placed on their specific locality by the rest of the organism in conjunction with its external experience via the central mind. Because physiological processes are derived from subjective intelligent interaction at a given level and locality, many of their aspects may change, depending on perceptions of the conditions at any given internal location. A given process or structure will not change just because of a given condition that exists right now; it will consult its subjective historical knowledge and make decisions on the basis of past experiences as well as current input.

OPPORTUNITY OF INPUT DURING DEVELOPMENT

Development is very similar to and tends to mirror evolution in its process, and development has been used in the understanding of evolution in some fields of biology. Early in development, all multicellular organisms start as a single cell, then become groups of cells, which are constantly dividing and migrating, forming the layers discussed above. At early enough developmental stages, it has been shown that a cell can become most any cell type, depending on the environmental signals it receives—that is, where it is located during development. Cells that would normally become a liver cell, for example, can become an entirely different type of cell if they are relocated early enough in development. This phenomenon is important in understanding mind–body medicine: It is the cell's perception of its local environment, not just its set of instructions, that induces its development and that affects the cell throughout its life. The basic idea of a stem cell is that it can theoretically become most any type of cell, in that it has not *differentiated*, or specialized into a specific cell type and function. Each stem cell has the ability to read a new environment and change its individual physiology to become whatever type of cell makes sense on the basis of its location or environment. When we say it becomes the cell type that *makes sense*, we mean that it purposely cooperates locally in a mutually beneficial way.

This is obviously very valuable for use in healing. Theoretically, a stem cell placed in the heart would become a heart cell; in the brain, a brain cell; and so on. Interestingly, and perhaps more importantly it has recently been found that there are far more stem cells that remain in the body as we age than was previously thought.

Furthermore, and more to the point of mind–body medicine, many cells have the ability to turn back into stem cells or, at least, to become more flexible (i.e., to *dedifferentiate*), given the right signals to do so. This is not unlike what happens after a bone break, as was described earlier. Some cells are flexible enough to change form or migrate in order to repair the bone. The point here is that natural healing processes are intelligent and induced subjectively by local conditions, depending on the cell's internal position and history. Furthermore, included in the local conditions is the influence of the central mind—again, as subjectively perceived by the local minds.

SYNCHRONIZATION OF LEVELS OF INTELLIGENCE

All living cells are intelligent and are aware of their respective environments, interpreting cues for what to do and how best execute their role in a given cooperative group.

Within the body, there are many examples of pathologies in which cells change into entirely different types of cells when their environment changes. According to intended evolution, this ability to change, repair, or remodel could be called the body's own internal medicine and is a basic mechanism of mind–body medicine. A relatively healthy body—and its cells and systems—already knows what to do in relation to the original historical central plan, given the correct conditions. Built-in flexibility capacity responds to control a return to baseline, for example, whereas pathology can be viewed as being away from baseline or a lack of the control at a given level or locality that is needed to return to it.

This is not to say that the body will heal anything on its own or with the help of the mind; we do not yet know the extent to which mind–body medicine can be used. The flexibility and healing capability of various functional groups or organs needs to be explored to identify potential targets for the use and development of virtual environments that can induce positive changes. For example, we might guess that using the (central) mind to create change within an automatic process would be very difficult, if it were even possible, and yet, we know that we can easily stimulate automatic processes in this way.

Improving the environment relevant to an unhealthy state where possible may be useful in accompanying other treatments when talking about curing disease. Current treatments often include physical repairs or blocking a function to stop a given symptom. Augmentation of this type of intervention by the mind could be used as an ally in helping a given unit or system repair itself or restore its function. Researching the potential to affect inner environments through both physical and mental activity could yield important insights in many biology-related fields, including medicine.

INFORMATIONAL STRESS AS A CONTRIBUTING FACTOR OF DISEASE

Although, given the theoretical view of intended evolution, we don't really differentiate between physical and mental modes, we will talk here about the common term *mind* in relation to disease. Probably the most obvious example of how important the mind is in health and medicine is the well-studied—if not well-understood—link between physical disease and mental stress. By our framework, when people talk about stress in a pathological way, it is said to occur when perceptions create an internal state that challenges the flexibility of part of the internal system such that an inordinate amount of time is needed to return to the baseline state: Therefore, stress occurs when environmental demand pushes the limits of built-in flexibility. Basically, according to the intended evolution framework, this occurs when perceived information is getting cued up in the processing portion of the information cycle but not being cleared in a timely manner. Reasons for this could include mismatched time lines of incoming demands and existing solutions or simply too much information of any kind to be processed at once. Another example of this is when there are internal deficiencies or disease states, which inhibit physical change, creating a backlog in local information cycles and also reducing the ability of the central cycle to deal with current incoming information. The important point is that what we call *stress* is the individual's perception and interpretation of the environment being passed to the internal systems, which also react subjectively to the information.

Perception—or, we could say, the mind—therefore, is the interface between the internal tiers and the external environment. That said, the mind that perceives the stressful information is actually composed of the collective minds of the internal systems, which are involved in interpretation and feedback to the central mind. In humans, this means that any perception—even an internally generated one—that is held in the mind becomes part of the internal environment that the inner systems subjectively perceive, process, and act on. Essentially, the central mind is part of the internal environment that the cells, tissues, and larger systems of the body react to. Perception, therefore, is an important input and inducer of physiological activity, including remodeling.

When we look at the health of the human body, different problems will manifest on the basis of evolutionary history and may be intelligent responses to the affected area's environmental input. In the human body, our mind functions as a sort of project manager and has the ability both to supply input to any area of the body and to receive feedback from those same areas. This multidirectional experience of perception means that intelligence and decisions or actions are subjective according to their locality; different internal functional groups read the same incoming signal in their own ways. Different internal functional groups also communicate their information back to the mind in their own subjective ways.

Finally, from the very beginning of the book, we know that perceptions of the external environment by the central mind are based on internal states, and therefore, any imbalances or problems affect the central perceptive experience.

The result of the foregoing is that the human mind can also create (virtual) environments for the inner tiers of the body, not just relay information to and from the body. The internal systems of the body react to the mind's perceptions, whether they are from the external environment or generated by the mind. They are also subject to learning—updating their knowledge—through virtual environments in the same way that all of life interacts and learns from its environment. If you close your eyes and imagine biting into a lemon, your physiology responds, regardless of the fact that you did not really bite a lemon. Sports psychology has many examples of training the mind, such as when a golfer imagines the perfect shot before hitting it. This type of virtual environment augments actual muscle memory and affects all of the systems down to the individual cells.

POTENTIAL HEALTH BENEFITS OF MIND—BODY MEDICINE

In line with currently accepted science showing stress to be a factor in disease states, mind–body medicine has successfully used relaxation and calming practices such as yoga and *qigong*. This makes perfect sense within the context of intended evolution, in that it allows the processing portion of the information cycle to be cleared or to be focused on information that is more easily processed internally. This is obviously beneficial for the purpose of augmenting other therapies or as a preventative technique, in that any stressed system has limited flexibility to heal local problems if it is already overwhelmed by signals from the central mind. Stress is being found to coincide with many pathologies, yet is difficult to quantify because people tend to handle it internally in differing ways. Scientifically validated mind–body medicine appears to be primarily based on identifying stress and attempting to reduce it.

These facts imply that much more is possible. Over time, repeated visualizations, together with exercises or movement, can train the mind and body to attain a higher level of health and well being, not just a better golf shot. By running a virtual environment in your mind, in combination with physical activity, one can "feed" the body and mind beneficial patterns of information, creating an environment more conducive to projecting more effectively what challenges lie ahead. This improves the "soft" flexibility described earlier, because planning ahead automatically reduces surprises and stress.

This is important in moving forward in mind–body medicine, because it means that we can be proactive in changing the health of the body. By using our minds to create virtual environments in conjunction with physical activity conducive to correcting problems, preventing new ones, or even increasing longevity, we can induce the body to act in a healthier way.

Creating a proactive virtual environment (like any lifeform interfacing with its environment) recontextualizes previously learned patterns of information. This can create longer lasting change than simply blocking out stressors for short periods of time or taking a vacation more often. A recontextualization of a previous pattern, done correctly, can lead to the emergence of a new healthier pattern; the original pattern is changed and the subjective stressor is recognized differently and can lessen, eliminate, or be avoided in the future. A virtual environment that includes longer time lines will tend to make incoming information that is short-term in nature more easily cleared and viewed subjectively in the context of the longer term.

The usefulness of current widely used techniques for relaxation are not being diminished here; in fact, they could provide the base structure for more advanced techniques that we are hoping this book can help future inventors develop. We want to also briefly note here that most standard physical exercise is also effective, in the sense that it works on the other end of the information cycle: the action portion. Physical activity can help clear the processing portion from that end, especially by doing an activity that reduces mental processing of the previous backlog (e.g., thinking about problems). This is also effective and obviously beneficial, and rather than being downplayed, we would hope that standard fitness and exercise professionals incorporate some of the framework presented here to help them make their programs even more effective.

Finally, the basic message to take away from this chapter is that common-sense approaches to good health must include an awareness that what we perceive programs our body on an ongoing basis. Perceptions are internalized and change us internally, including physically. Environments causing quick or stressful changes in emotional states or stress require quick action and physical adjustment—essentially requiring increased amounts of energy for constant homeostatic rebalancing, challenging our flexibility factor. We believe new healing methods can be discovered with thought, research, and experimentation. Mind–body medicine is in its infancy, and it is our hope that this book is helpful in furthering its development.

CONCLUSION

Simple processes often produce great complexity. The complexity of life is a reflection of the environment that holds it, but the process of life itself is simple. This process takes place at the information level, is experienced and unfolds internally, and reaches far beyond a simple survival-of-the-fittest paradigm. Many times, some critical aspect of survival takes place purely in the realm of possibilities and not in actual, physical struggles. When we talk about life's journey on Earth, we should look at it as a journey taken with self-purpose, a series of self-propelling actions that lead toward a common direction. Life is about change, because all things change; its overall direction comes from the basic intention that is driving it forward as guided by subjective intentions.

Information lies at the heart of the living process. Life, above all else, is informational in nature, and we believe this framework can help explain how life has evolved as it moved forward in time. Survival was never the end goal for life; rather, it is what allowed growth and expansion, not just physically but informationally as well. Human evolution is at a point at which survival pressure is a much smaller factor than ever before. We can make intelligent decisions about our own future—not just about how we can survive but also about how we can thrive. We can do this by taking advantage of the tools built inside of us through countless rounds of knowledge collection and interface with our environments. By simply and efficiently meeting our own needs, we can intelligently create our future.

So the question is not just what can we achieve, but more importantly, what we want to achieve. We must then optimize our mind's intelligence, our body's functioning, and our mastery of technology to make that intention come to pass.

ABOUT THE AUTHORS

—

Dongxun Zhang is the creator of both the theory of intended evolution and the Intended Evolution Fitness System. Zhang is a doctor of acupuncture and oriental medicine (DAOM). He is a professor of the doctoral program of Texas Health and Science University and on the faculty of Consilient Innovation Network, and a former director of the International Association of Integrated Medicine, for whom he has addressed the United Nations. In 1997, he was recognized by the Sixth International Traditional Chinese Medicine Conference with the Hwang-Di Award for his microdiagnostic system.

Bob Zhang was born in China and moved to the United States at the age of ten. He went to middle school, high school, and college in Austin, Texas. He is married and has two children.

Editor **David Kincade** holds BS degrees in biology and economics and an MS in Oriental Medicine (MSOM). He practices Oriental Medicine in Austin, Texas.